ASSESSMENT OF PROGRAMS IN SPACE BIOLOGY AND MEDICINE 1991

COMMITTEE ON SPACE BIOLOGY AND MEDICINE

L. Dennis Smith, University of California, Irvine, *Chairman*
Robert M. Berne, University of Virginia, Charlottesville
Peter Dews, Harvard Medical School
R.J. Michael Fry, Oak Ridge National Laboratory
Edward Goetzl, University of California Medical Center, San Francisco
Robert Helmreich, University of Texas, Austin
Barry W. Peterson, Northwestern University
Clinton T. Rubin, State University of New York, Stony Brook
Alan L. Schiller, Mt. Sinai Medical Center
Tom Scott, University of North Carolina, Chapel Hill
William Thompson, North Carolina State University
Fred W. Turek, Northwestern University

Staff

Joyce M. Purcell, Senior Program Officer
Melanie M. Green, Administrative Secretary

SPACE STUDIES BOARD

Louis J. Lanzerotti, AT&T Bell Laboratories, *Chairman*
Philip H. Abelson, American Association for the Advancement of Science
Joseph A. Burns, Cornell University
John R. Carruthers, INTEL
Andrea K. Dupree, Center for Astrophysics
John A. Dutton, Pennsylvania State University
Larry W. Esposito, University of Colorado, Boulder
James P. Ferris, Rensselaer Polytechnic Institute
Herbert Friedman, Naval Research Laboratory (retired)
Richard L. Garwin, IBM T.J. Watson Research Center
Riccardo Giacconi, Space Telescope Science Institute
Noel W. Hinners, Martin Marietta Corporation
James R. Houck, Cornell University
David A. Landgrebe, Purdue University
Elliott C. Levinthal, Stanford University
William J. Merrell, Jr., Texas A&M University
Richard K. Moore, University of Kansas
Robert H. Moser, The NutraSweet Company
Norman F. Ness, University of Delaware
Marcia Neugebauer, Jet Propulsion Laboratory
Sally K. Ride, University of California, San Diego
Robert F. Sekerka, Carnegie Mellon University
Mark Settle, ARCO Oil and Gas Company
L. Dennis Smith, University of California, Irvine
Byron D. Tapley, University of Texas, Austin
Arthur B.C. Walker, Stanford University

Marc S. Allen, Staff Director

COMMISSION ON PHYSICAL SCIENCES, MATHEMATICS, AND APPLICATIONS

Norman Hackerman, Robert A. Welch Foundation, *Chairman*
Peter J. Bickel, University of California at Berkeley
George F. Carrier, Harvard University
Herbert D. Doan, The Dow Chemical Company (retired)
Dean E. Eastman, IBM T.J. Watson Research Center
Marye Anne Fox, University of Texas
Phillip A. Griffiths, Duke University
Neal F. Lane, Rice University
Robert W. Lucky, AT&T Bell Laboratories
Christopher F. McKee, University of California at Berkeley
Richard S. Nicholson, American Association for the Advancement of Science
Jeremiah P. Ostriker, Princeton University Observatory
Alan Schriesheim, Argonne National Laboratory
Roy F. Schwitters, Superconducting Super Collider Laboratory
Kenneth G. Wilson, Ohio State University

Norman Metzger, Executive Director

Foreword

This report is one in a series written by the standing discipline committees of the Space Studies Board. The purpose of this new series is to assess the status of our nation's space science and applications research programs and to review the responses of the National Aeronautics and Space Administration and other relevant federal agencies to the Board's past recommendations.

It is important, periodically, to take stock of where research disciplines stand. As an advisory body to government, the Space Studies Board should regularly examine the advice it has provided in order to determine its relevance and effectiveness. As a representative of the community of individuals actively engaged in space research and its many applications, the Board has an abiding interest in evaluating the nation's accomplishments and setbacks in space.

In some cases, recurring budget problems and unexpected hardware failures have delayed or otherwise hindered the attainment of recommended objectives. In other cases, space scientists and engineers have achieved outstanding discoveries and new understandings of the Earth, the solar system, and the universe. Although the recent past has seen substantial progress in the nation's civil space program, much remains to be done.

These reports cover the areas of earth science and applications, solar system exploration (and the origins of life), solar and space physics, and space biology and medicine. Where appropriate, these reports also include the status of data management recommendations set forth in the reports of the Space Studies Board's former Committee on Data Management and Computation. The Board has chosen not to assess two major space research disciplines—astronomy and astrophysics, and microgravity research—at this time. Astronomy and astrophysics was recently surveyed in a report under the aegis of the Board on Physics and Astronomy, *The Decade of Discovery in Astronomy and Astrophysics* (National Academy Press, Washington, D.C., 1991); the Space Studies Board is currently developing a strategy for the new area of microgravity research.

On completion of the four reports, the Board will summarize the contents of each volume and produce an overview. The Space Studies Board expects to repeat this assessment process approximately every three years, not only for the general benefit of our nation's space research programs, but also to assist the Board in determining the need for updating or revising its research strategies and recommendations.

Louis J. Lanzerotti
Chairman, Space Studies Board

Contents

SUMMARY 1

1 INTRODUCTION 10

2 SCIENCE PROGRAM AND POLICY ISSUES 15

3 HUMAN PHYSIOLOGY 20

4 BEHAVIOR, PERFORMANCE, AND HUMAN FACTORS 35

5 DEVELOPMENTAL AND CELL BIOLOGY 40

6 PLANT BIOLOGY 47

7 CLOSED ECOLOGICAL LIFE SUPPORT SYSTEMS 52

8 RADIATION BIOLOGY 59

9 CONCLUSIONS 63

BIBLIOGRAPHY 67

APPENDIX: Guidelines for Assessment Reports for
Standing Committees of the Space Studies Board 69

Summary

INTRODUCTION

This report was undertaken at the request of the Space Studies Board to provide an up-to-date assessment of the status of the implementation in the civil space program of the various research strategies and recommendations published in previous reports. This report limits its comments to information contained in the three most recent reports (SSB 1979, 1987, and 1988). The most comprehensive strategy was the report published in 1987, *A Strategy for Space Biology and Medical Science for the 1980s and 1990s* (SSB, 1987), edited by Jay Goldberg, University of Chicago. The Goldberg Strategy (as the 1987 strategy report is referred to in this report) forms the primary basis for the current evaluation, although reference is also made to several previous reports concerning life sciences by the Committee on Space Biology and Medicine (CSBM) and the Life Sciences Task Group of the Space Science Board that was part of the 1988 *Space Science in the Twenty-First Century* report. Space biology and medicine includes—in addition to biological and medical subdisciplines—human behavior, radiation, and closed ecological life support systems.

The Goldberg Strategy defined four major goals:

1. To describe and understand human adaptation to the space environment and readaptation upon return to earth.

2. To use the knowledge so obtained to devise procedures that will improve the health, safety, comfort, and performance of the astronauts.

3. To understand the role that gravity plays in the biological processes of both plants and animals.

4. To determine if any biological phenomenon that arises in an

individual organism or small group of organisms is better studied in space than on earth.

The first two goals have taken on new emphasis since the announcement of the Space Exploration Initiative (SEI), enunciated by the President in July 1989, for a sequential progression of human activities in space, and extending potential human missions to years in duration. In discussing the major imperatives for research in space biology and medicine, this assessment of the implementation of the research strategies has categorized research topics relative to the urgency that would be dictated by proceeding with a space exploration initiative.

The conduct of research in space biology and medicine is influenced by the way the civil space agency, National Aeronautics and Space Administration (NASA), is structured and managed. Consequently, previous reports by CSBM have contained numerous recommendations concerning science program and policy issues. Because of the importance of these issues in approaching the various research goals, progress in implementation of these goals is discussed at the outset. This is followed by topics that have the greatest potential of affecting human performance and/or survivability during sustained space exploration. These topics include research areas concerning human physiology in microgravity, human behavior during long-term missions, and the radiation environments of space. Finally, the report contains sections on developmental and cell biology, human reproduction, plant biology, and issues associated with the development of a closed ecological life support system. The latter topics reflect areas that, while not deemed crucial to survival in space for durations of a few years, could become critical to longer term human habitation. In addition, these topics represent major research areas in which space could be especially valuable in the study of basic biological phenomena.

SCIENCE PROGRAM AND POLICY ISSUES

Published strategy reports (e.g., SSB 1979, 1987, 1988) contain recommendations concerning how NASA manages its life sciences research and the design and utilization of laboratory space on a space station. In the area of management, recommendations were as follows:

1. Standing panels of 5 to 10 scientists be created to review, update, and refine research strategies in each subdiscipline of space biology and medicine.

2. Announcements of Opportunity (AOs) and NASA Research Announcements (NRAs) concerned with Shuttle flights and the space station should be targeted to a particular subdiscipline and should

state explicitly the major research questions that the mission is intended to address.

3. NASA should actively solicit the participation of other relevant federal agencies such as NIH and NSF in the design and conduct of research related to the major questions that need to be answered.

Recommendations related to a space station were as follows:

1. A space station should contain a dedicated life sciences laboratory, and research time should be allocated in 3- to 6-month increments for individual subdisciplines.

2. A variable force centrifuge of the largest possible dimensions should be incorporated into a space station.

3. Dedicated microprocessors should be used for process control, data storage, or both, and rapid communication in real time with ground-based research teams should be a goal.

In the area of management, NASA either has implemented or is in the process of implementing all of the recommendations made. The internal life sciences advisory has been reorganized as recommended. The NRAs that are now being released are more highly focused, and NASA is now actively cooperating with other federal agencies such as National Institutes of Health (NIH), National Science Foundation (NSF), and the U.S. Department of Agriculture (USDA), as well as numerous foreign partners.

None of the recommendations concerning design and utilization of the space station have been implemented in current plans for the facility; however, planning for inclusion of a centrifuge is under way.

RESEARCH IN SPACE BIOLOGY AND MEDICINE

Human Physiology

There has been a general perception that since a small number of Soviet cosmonauts have survived in the microgravity of space in low earth orbit for as long as a year, there are no major physiological problems likely to preclude longer term human exploration beyond low earth orbit. The committee has had, over the years, access to anecdotal data from the Soviet space program. This anecdotal information is, while interesting, not sufficiently reliable for drawing conclusions or in planning the U.S. program for a number of reasons. There are differences in experimental protocols and controls in laboratory equipment, and the Soviets do not publish their results in refereed scientific journals. However, increased recent cooperative activities between the Soviets and the United States suggest promise for the future in standardized experimental procedures and data exchange.

The current evaluation of progress in space biology and medical research illustrates that all of the major physiological problems characteristic of prolonged human exposure to the microgravity environment of space remain unsolved. First, and of greatest concern, is bone, muscle, and mineral metabolism; second, cardiovascular and homeostatic functions; and third, sensorimotor integration.

Bone, Muscle, and Mineral Metabolism

Eight major goals were defined for the study and bone and mineral metabolism: (1) determine the temporal sequence of bone remodeling in response to microgravity; (2) establish the reversibility of this process on return to a 1-g environment; (3) establish the relationship between muscle activity and bone function; (4) devise countermeasures to prevent bone loss; (5) establish the cellular mechanisms responsible for bone loss; (6) evaluate the interdependence of calcium homeostasis and bone remodeling; (7) determine the etiology of pathologic calcification; and (8) establish the biomechanics of the skeleton under microgravity conditions.

Understanding the etiology of bone loss (osteopenia) is the focus of an enormous research program within the NIH as well as an area of research that has received major attention by NASA scientists—especially over the past 5 years. NASA scientists and others supported by NASA have developed an animal model to study bone loss. In addition, human studies correlating inactivity (bed rest) to factors such as diminished bone mass and increased urinary calcium have also proven to be useful models for potential changes during extended spaceflight. However, of the eight major goals listed above, only the first has been addressed in these studies, and the information that has been obtained using the animal model chosen (rat) is of limited value because of the dissimilarities between bone physiology in rats and normal human physiology. Considerable research remains to be conducted. Increased interaction with the major research effort at NIH would be of enormous value for solving the overall problems of bone and muscle atrophy that have been observed in microgravity.

Cardiovascular and Other Homeostatic Systems

The cardiovascular and neuroendocrine elements of the circulatory system focus respectively on basic cardiovascular function and the influences of regulatory systems on these functions. Additional areas under this topic include immunology, hematopoiesis, and wound healing.

Circulatory Adjustments The major goals have been to (1) understand acute (0 to 2 weeks), medium-term (2 weeks to 3 months), and long-term (greater than 3 months) changes in the cardiovascular system in microgravity; (2) examine the validity of ground-based models of microgravity-induced changes; and (3) define measures (countermeasures) that will alleviate changes in microgravity and hasten human adaptation upon return to a 1-g environment.

A better understanding of cardiovascular and pulmonary physiology in microgravity has been a major goal of previous, current, and planned investigations. Measurements on humans before, during, and after several Shuttle flights have provided echocardiographic data on cardiac dimensions and function. Some countermeasures such as oral saline loading have been tested to prevent post-flight orthostatic hypotension. A major drawback has been the limited number of subjects available for study. There is a need to develop animal models for both ground-based and flight experiments. Hormones that affect the cardiovascular system also remain to be tested in the context of cardiovascular changes that occur in space. Some hormone measurements were conducted on Skylab flights, and additional studies are planned on upcoming Shuttle flights. However, many of these experiments fail to take into account fairly recent observations concerning the rhythmic nature of changes as a function of circadian variations.

Immunology Immune cells in mammalian bone marrow and lymphoid organs initiate and regulate lymphocyte and antibody responses as well as control the production and function of cells in the blood and connective tissues. The major goal in this area is to determine if cells of the immune system can proliferate in space and maintain a normal immune system.

The occurrence of serious infections in space has been very uncommon, and most studies of immunity in space have been directed to the detection of abnormalities in human and animal lymphocyte numbers and morphology in space. Spaceflight is known to result in significant reductions of both plasma volume and red blood cell mass within days. Recent studies have shown that lymphocytes do not respond to stimuli that normally cause division, suggesting an impaired ability to proliferate in space. This could have profound implications to the immune and hematopoietic system. An expanded effort to investigate possible immune deficiencies coupled with the development of cell models to test immune and bone cell function in microgravity requires a higher priority.

Sensorimotor Integration

As indicated in the Goldberg report, the neuronal mechanisms underlying a sense of spatial orientation are complex, as yet poorly understood, and are directly relevant to assuring the effective functioning of humans involved in space missions. The 1987 strategy report recommended a vigorous program of ground-based and flight research aimed at understanding these mechanisms as they operate on earth, in space, and on return from microgravity to high-gravity environments. These studies become all the more significant if one considers the use of artificial gravity (rotating spacecraft) as an attempt to ameliorate the effects of microgravity on human physiology. Specific goals are to (1) study in microgravity how the vestibuloocular reflex (VOR) converts head motion into compensatory eye movement, (2) investigate the neural processing mechanisms in the vestibular system in both normal gravity and microgravity, (3) focus on adaptive mechanisms that alter vestibular processing in response to altered feedback from the environment, and (4) investigate more fully the etiology of motion sickness in microgravity.

Overall, NASA has made a concerted effort to undertake appropriate, quality research in the sensorimotor area. These efforts include many studies supported through external investigators and the establishment of an excellent Vestibular Research Facility (VRF) at Ames Research Center. In spite of limited flight opportunities, considerable progress has also been made studying sensorimotor performance in microgravity. Several planned experiments are promising. However, in spite of this generally positive view, no single countermeasure has yet been developed that corrects the problem of space motion sickness. Perhaps the syndrome, with individual variations, is actually several distinguishable syndromes. This possibility, if documented, might dictate new research approaches.

Behavior, Performance, and Human Factors

The major goals for space research as it relates to human behavior are to develop (1) spacecraft environments, (2) interfaces with equipment, (3) work-leisure schedules, and (4) the social organization that will optimize the efficiency, safety, and satisfaction of crews during long-term spaceflight.

With the exception of group and organizational factors, there is research in progress along the lines recommended in published research strategies. Much of the progress that has occurred derives from well-funded research programs in aviation sponsored by the Federal Aviation Administration (FAA) and to a lesser extent from NASA's aviation research program. However, this type of research, while useful, cannot provide all of the information needed to support a long-term human presence in space. As opportunities for experimentation that will exist during long-duration spaceflight will always be extremely limited, there must be a well-developed ground-based program of research employing a variety of research settings. At this point in time, NASA has no plans to develop long-term confinement studies using ground-based research settings.

Developmental and Cell Biology

The major goal for developmental biology as outlined in all three research strategies is to determine whether any organism can develop from fertilization through the formation of viable gametes in the next generation, i.e., from egg to egg, in the microgravity environment of space. In the event that normal development does not occur, the priority is to determine which period of development is most sensitive to microgravity. Potentially, research on specific developmental phases (e.g., fertilization to initial organ formation) would suggest detailed studies on the function and differentiation of individual cells or groups of cells. In approaching these goals, we have recommended studies on several representative organisms including both invertebrate and vertebrate animals. While the latter would include mammals such as mice, it also encompasses the question, can humans reproduce in space? The importance of these questions relates to the ability to establish permanent human colonies in space as well as to the possibility that the space environment could be a particularly advantageous environment to study basic developmental research.

A number of diverse organisms have been subjected to microgravity for varying periods of time. The results of these studies have been inconsistent. Both normal and abnormal development have been observed, dependent on the organism and the stage of development at which the material was subjected to microgravity. To our knowledge, no animal species has ever been carried through one complete life cycle in the microgravity of space.

Plant Biology

Any strategy that visualizes a long-term sustained human presence in space absolutely requires the ability to continuously grow and reproduce various plant species over multiple generations. A related goal, which has implications for agriculture generally, is to understand the mechanism(s) involved in gravity sensing by plants. This requires an emphasis on ground-based research as well as research in space.

For the most part, observations on plants exposed to microgravity have been anecdotal. It has been demonstrated repeatedly that plants do grow in microgravity. However, whether plants can grow normally remains to be determined. Significantly, results of studies on the German D-1 mission, which incorporated onboard 1-g centrifuge controls, indicate that single plant cells behave normally or even exhibit accelerated development. In contrast, the roots of seedlings germinated in microgravity grew straight out from the seed, and the same roots contained starch grains (statolyths) which were more or less randomly distributed in their cells. Control roots centrifuged at 1 g on the flight, were normally gravitropic.

Cytological studies of roots flown under a variety of conditions in space have consistently revealed reduced cell divisions as well as a variety of chromosomal abnormalities. At the same time, some Soviet experiments using the plant arabdiposis indicate that at least this plant develops normally through the flowering stage. However, in the Soviet experiments, fruit set was decreased and seeds brought back to earth germinated less efficiently than ground based controls. Long-term flight experiments are required to determine if a variety of plant species can grow normally in microgravity and, in particular, if they can produce viable seeds.

Closed Ecological Life Support Systems

The closed ecological life support system (CELSS) program at NASA is attempting to create an integrated self-sustaining system capable of providing food, potable water, and a breathable atmosphere for space crews during missions of long-term duration. An effective CELSS must have subsystems both for plant and animal growth, food processing, and waste management. These have been described to some extent on previous pages. A CELSS must be much more than a "greenhouse in space." It must be a multispecific ecosystem operating in a small closed environment. Thus, although the concept is easily articulated, numerous areas of ignorance remain.

Based on consideration of primarily agricultural plant species, a small number have been selected for further investigation. These include wheat, potato, soybean, and tomato. Growth chamber studies have been initiated, both at NASA and in university laboratories, with the aim of defining the conditions required for optimum rates of dry matter production. Although most research has been done with open systems, experiments with closed systems have recently been initiated. No attention has been paid to the use of techniques of plant breeding or genetic engineering to "design" ideal plants for a CELSS system. No experiments have yet been performed in microgravity to determine if current systems can function in space. In short, a considerable increase in research efforts, and support for those efforts, is required in order to reach the desired goals.

Radiation Biology

While the radiation environment within the magnetosphere is fairly well known, as are the biological effects of low energy transfer (LET) radiations from protons and electrons, considerably better quantitative data on LET dose rates beyond the magnetosphere are still required. In particular, better predictability of the occurrence and magnitude of energetic particles from solar flares is required; radiation from solar flares can be life-threatening in relatively short time periods. Major goals of radiation research are to quantify high-energy (HZE) particles in space and to understand the biological effects of HZE particles. The likely long-term biological effects of exposure to HZE particles is an increased incidence of cancer and brain damage.

NASA has maintained a limited but ongoing research program both in radiation dosimetry and radiobiology including ground-based programs on the effect of fragmentation of HZE particles and on the secondary particles. In the field of radiobiology, NASA has supported studies dealing with the biological effects of HZE particles. Limited flight data suggest a synergism between HZE particle hits and microgravity. This research requires increased attention. In particular, ground-based studies on biological effects of HZE particles are currently performed in the United States at the Billion Electron Volts Linear Accelerator (BEVALAC) at Lawrence Berkeley Laboratory. This research may be drastically curtailed if the facility is unavailable after 1993 as is currently planned. Use of similar facilities in other countries, while feasible, is not necessarily practical because of the necessity for transporting large numbers of animals and associated experimental controls, and regular transport and accommodation of U.S. research teams.

CONCLUSIONS

Over the past 30 or more years, the Space Studies Board and its various committees have published hundreds of recommendations concerning life sciences research. Several particularly noteworthy themes appear consistently: (1) *balance*—the need for a well-balanced research program in terms of ground versus flight, basic versus clinical, and internal versus extramural; (2) *excellence*—because of the extremely limited number of flight opportunities (as well as their associated relative costs), the need for absolute excellence in the research that is conducted, in terms of topic, protocol, and investigator, and (3) *facilities*—the single most important facility for life sciences research in space, an on-board, variable force centrifuge.

In this first assessment report, the Committee on Space Biology and Medicine emphasizes that these long-standing themes remain as essential today as when first articulated. On the brink of the twenty-first century, the nation is contemplating the goal of human space exploration; consequently, the themes bear repeating. Each is a critical component of what will be necessary to successfully achieve such a goal.

1

Introduction

AREAS/DISCIPLINES OF ADVISORY RESPONSIBILITY

Space biology and medicine studies how individual organisms and small groups of organisms respond to the microgravity of space and how they adapt to it. It has been clear for sometime that when humans go into space, many changes occur in their physiology. Several studies have also indicated that basic biological processes are altered in microgravity. It is not likely that the processes are separable. Human physiology is predicated on the homeostatic functioning of organs that are composed of cells. All of these complex functioning systems have evolved in the presence of gravity, and when exposed to microgravity, they are forced to function in a new and novel environment. Thus, in attempting to understand the adaptation to microgravity, scientists are forced to evaluate not only the clinical manifestations of an organismal response to the new environment but also the underlying cellular and organ response. This requires an integrated approach that includes both basic research as well as the more operational aspects of clinical research.

One approach to understanding adaptation to microgravity involves empirical research in which humans or appropriate animal models are subjected to the space environment for prolonged periods and are monitored continuously for signs of changes. This process might lead to putative countermeasures that may provide a "fix" for the problems encountered, but it is not likely to elucidate the basic mechanism(s) involved in the response to microgravity. A more appropriate research strategy is to study basic mechanisms and, based on the knowledge acquired, to design appropriate countermeasures. This is a longer-term process. In practice, both approaches require access to research facilities in space and on the ground.

The Committee on Space Biology and Medicine (CSBM) and several of its predecessors have formulated comprehensive research strategies for understanding both basic and clinical aspects of adaptation to microgravity (see below). The Space Exploration Initiative (SEI), enunciated by the President in July 1989, envisions a sequential progression of human activities in space of many years' duration. This has placed increased emphasis on implementation of the appropriate research strategies. Ironically, since a small number of Soviet astronauts have survived in the microgravity environment of space *in low earth orbit* for as long as a year, the perception has developed that there are no major physiological or psychological problems likely to preclude longer-term human exploration beyond low earth orbit. The fallacy of that assumption has been documented in previous reports and the current document reaffirms that conclusion.

The current report was undertaken to provide an up-to-date evaluation of the extent to which National Aeronautics and Space Administration (NASA) has implemented the various research strategies published over the past 10 years. Since 1963, various forms of the current CSBM have existed. The evolution of the advisory responsibilities and membership of the committee has reflected the times, paralleling the needs of NASA and the U.S. Space Program. Over this nearly 30-year period, the committee or its predecessors have issued, or contributed to, nearly 20 reports (see the bibliography) on space biology and medicine topics, some narrowly focused (e.g., *Report of the Panel on Management of Spacecraft Solid and Liquid Wastes*, SSB, 1969a), others covering the whole of space life sciences (*Space Science in the Twenty-First Century: Life Sciences*, SSB, 1988). Still other reports and official letter reports to NASA concern matters more of a policy or programmatic nature (*Life Sciences in Space*, SSB, 1970c; letter to Associate Administrator Stofan, NASA, regarding centrifuge, July 21, 1987; letter to Administrator Truly, NASA, regarding the extended duration orbiter medical program, December 20, 1989). This report limits its comments to information contained in the three most recent reports (SSB 1979, 1987, and 1988).

The predecessor to the 1987 strategy report was *Life Beyond the Earth's Environment* (SSB, 1979). This report was not a formal Space Studies Board strategy *per se*. However, it covers virtually all of the major space biology and medicine disciplines and makes recommendations for conducting research in each of these fields. It addresses several areas not individually treated in the 1987 report, in particular, the effects of radiation on living organisms in space and closed ecological life support systems (CELSS). In 1988, SSB published the results of a three-year study that included individual task group reports for each major space science discipline, including life sciences (i.e., *Space Science in the Twenty-First Century: Imperatives for the Decades 1995 to 2015*). The Life Sciences Task Group report addresses all of space life sciences, including exobiology, global biology,

CELSS, instrumentation and technological requirements, and space biology and medicine. The CSBM's advisory purview is limited to this report's recommendations in space biology and medicine, CELSS, and relevant instrumentation and technology requirements.

A number of reports have been issued by the SSB's Committee on Data Management and Computation (CODMAC): *Data Management and Computation—Volume 1: Issues and Recommendations* (SSB, 1982), *Issues and Recommendations Associated With Distributed Computation and Data Management Systems for the Space Sciences* (SSB, 1986), and *Selected Issues in Space Science Data Management and Computation* (SSB, 1988). CODMAC also assumed the responsibility for its recommendations concerning data issues in the life sciences.

The most recent report of the CSBM was A *Strategy for Space Biology and Medical Science* for the 1980s and 1990s (SSB, 1987): Jay Goldberg, University of Chicago, chaired the CSBM that wrote the report. This report, hereafter referred to as the Goldberg Strategy, was transmitted to NASA in the spring of 1987. Since that time, over 3,500 copies have been distributed. The report is one of a number of scientific strategies that have been produced by SSB standing committees. It is meant to provide NASA with a guideline for developing its long-term mission plans and a rational, coherent research program in space biology and medicine.

PRINCIPAL USERS/IMPLEMENTORS OF DATA ON SPACE BIOLOGY AND MEDICINE

Within NASA

While the primary audience of the CSBM reports that have been published is the Life Sciences Division at NASA, the nature of space biology and medicine requires that all of NASA be considered to be the "user" of their recommendations. The operational requirements of maintaining the health and safety of humans in space both in the past and in the future are the focus and responsibility of offices in NASA in addition to the Life Sciences Division and the Office of Space Science and Applications. This has been true from the days of Apollo to the present days of the Space Shuttle. As NASA and the nation proceed to plan for implementing the SEI, the need for a comprehensive agency-wide understanding of the integrated, critical issues associated with a soundly based space biology and medical research program will be even more crucial.

Outside NASA

One of the principal agencies outside of NASA that is a potential user of CSBM reports/recommendations is the National Institutes of Health (NIH).

The committee encouraged collaborative efforts between NASA and NIH in the 1987 report. In this time of ever-increasing budgetary constraints, there is a need to maximize the scientific return from research in space biology and medicine. Collaborative activities between the two agencies to the fullest extent possible are strongly encouraged.

The committee is aware that NASA has begun discussions with the Department of Agriculture (USDA) concerning collaborations on research related to the CELSS program, molecular research, and plant biology. This sort of collaboration has the potential of enhancing these programs, which have been constrained by NASA over the years because of other, competing funding priorities.

The National Science Foundation (NSF) funds a significant amount of biological research that could have some relevance to the space program. Again, it is hoped that NSF and the community supported by NSF are cognizant of the recommendations and scientific issues raised in CSBM reports and that some cooperative/collaborative activities commence.

Another potential user of CSBM's advice would be the Department of Defense (DOD). To date, the committee has had no interaction with the DOD and is not informed as to the extent of that department's needs or activities. However, given the major extent of DOD's space program both in the past and that anticipated for the future, it is the committee's opinion that identifying areas of mutual interest and concern to the civilian and military space programs would be beneficial to both sectors. With the National Space Council now in place, perhaps this could be accomplished more easily.

To a somewhat lesser extent, NASA's human factors research would be of interest and complementary to certain activities of the Federal Aviation Administration (FAA). The FAA has the primary responsibility for implementing the 1988 Aviation Safety Research Act.

Finally, the committee hopes that policy makers from both the executive and legislative branches of government are aware of their advice and recommendations and their underlying rationales, particularly if announced plans for the SEI proceed.

PRINCIPAL RESEARCH STRATEGY

This report was undertaken to provide an up-to-date evaluation of the extent to which NASA has implemented the various research recommendations made in its various published reports. Since the Goldberg Strategy represents the most recent and comprehensive evolution of a research strategy, the CSBM has paid particular attention to the experimental approaches recommended therein. This is justifiable for a second reason. The Goldberg Strategy was written and published during the hiatus in flight opportunities resulting from the Challenger disaster. In the interim, Shuttle flights have resumed, and a comprehensive manifest exists of planned flights for the next several years. In terms of the previous strategies, it becomes all the more important to determine to what extent NASA's future plans include recommendations in that report.

It should be emphasized that the Goldberg Strategy and all previous reports referenced above deal with research strategies that assume a microgravity environment. Alternative strategies that involve the creation of an artificial gravity environment in space have not been dealt with and are not considered herein. The CSBM is aware of the need to address this specific issue (artificial gravity) in future deliberations.

The Goldberg Strategy contains 11 chapters—including one on developmental biology, one on gravitropism in plants, seven concerned with human physiology, one on human behavior, and one concerned with policy and programmatic issues.

In this report, recommendations on human physiology are discussed in one major chapter (Chapter 3) with subheadings. The section on gravitropism is incorporated into a chapter on plant biology (Chapter 6). This report also includes chapters on radiation biology (Chapter 8) and the CELSS program (7), which as mentioned earlier, were not part of the Goldberg Strategy, but were addressed in the 1988 and 1979 reports. Finally, recent developments in the area of cell biology are incorporated as a section in the chapter on developmental biology (Chapter 5).

2

Science Program and Policy Issues

INTRODUCTION

In reviewing the numerous reports published under its auspices, the Committee on Space Biology and Medicine (CSBM) was reminded of a long-standing consensus regarding the disciplines under its purview. A continuing and major feature of the committee's various deliberations concerns the manner in which NASA "does business" in the general area of life sciences. The nature of life sciences research requires a program more analogous to that of other federal agencies that support basic life sciences research than has been the case with NASA's life sciences program to date.

This is not to suggest that NASA should be reconfigured to function like the National Institutes of Health (NIH) or the National Science Foundation (NSF), or other relevant federal agencies. However, it has been suggested that since this type of research requires an empirical approach, a departure from those practices that have been traditionally employed in conducting NASA's space biology and medicine program may be justified and could prove more efficient for reaching its long-term goals.

The Goldberg Strategy contained several recommendations concerned with a number of science program and policy issues. Because of the perceived significance of these recommendations, they are discussed at the beginning of this evaluation.

MAJOR RECOMMENDATIONS

Science Program Issues

Space biology and medicine are dynamic sciences, continually evolving, sometimes quite rapidly, as new results and technological developments occur. Hence, the Goldberg Strategy recommended *that standing panels of 5 to 10 qualified scientists be created to review, update, and refine the research strategy in each subdiscipline of space biology and medicine.*

Since the facilities and personnel available for research activities in space biology and medicine, which are under the direct auspices of NASA, are extremely limited, and recognizing the importance of ground-based research, the committee recommended that NASA increase its interaction with the academic community.

As the design of experiments in space biology and medicine requires continuous access to space as well as the necessity for manned intervention, the Goldberg Strategy contained two major recommendations relative to the projected use of the Space Station. Specifically, the committee suggested that: *(a) there be a dedicated life sciences laboratory on the Space Station, i.e., research space on individual modules should not be shared with other disciplines, and (b) that space biology and medicine research time on the Space Station be allocated in 3- to 6-month increments, with each block devoted to a single research area such as neurovestibular research.* Related to this, *two or more payload specialists on each flight should be practicing laboratory scientists in the particular research areas assigned blocks of time on Space Station.*

Recognizing the need for controlled experiments as well as the need to evaluate the role of variable gravitational forces in affecting biological processes, the Goldberg Strategy (as well as numerous NASA and non-NASA reports) recommended in the strongest possible terms the requirement for a *variable force centrifuge of the largest possible dimensions in any facility designed to obtain data on the role of microgravity in affecting biological processes.*

Variable Force Centrifuge (VFC). Such a device is the single most important facility in any life sciences program. A VFC has three equally important functions. (1) It provides an on-board 1-g control that can separate the influence of weightlessness from the other effects of spaceflight. (2) Microgravity has both short-term and long-term effects on biological systems. Both kinds of effects involve important biological phenomena. Their study is greatly facilitated by a VFC, which allows exposure to microgravity or to gravitational forces for varying periods of time. (3) The removal of gravitational forces is already known to have major impacts on biological systems. In such cases, it is of particular importance to determine if there is a threshold force required for a response to occur and, more generally, to ascertain the dose-response relationship. A VFC offers the crucial advantage in answering these questions, since it makes possible the introduction of fractional g forces. From these comments, it can be appreciated that a VFC should increase the scientific return from space experiments by orders of magnitude.

This committee has been apprised of the engineering problems involved in the inclusion of a large centrifuge in a freely floating Space Station. Nevertheless, *the committee still recommends that a variable force centrifuge of the largest possible dimensions be designed, built, and included in the initial operating configuration of the Life Sciences Laboratory. A VFC is an essential instrument for the future of space biology and medicine.* (SSB, 1987, p. 15)

Finally, the committee provided several recommendations concerned with the acquisition, transmission, processing, and storage of data obtained from space. Specifically, it was recommended that *dedicated microcomputers and communications capabilities be used for process control, data storage, or both, and that rapid communication in real time with ground based research teams be a goal.*

Science Policy Issues

Because of the extremely limited access to space, the Goldberg Strategy recommended that *any Announcements of Opportunity (AOs) or NASA Research Announcements (NRAs) concerned with Shuttle flight or Space Station should be targeted to a particular subdiscipline and should state explicitly the major research questions that the mission is intended to address.*

While the responsibility for a coordinated program in space biology and medicine clearly rests with NASA, the facilities and personnel in NASA are insufficient to undertake the research necessary to address the biological problems associated with the effects of microgravity on biological processes. For this reason, the CSBM has strongly and repeatedly recommended *that NASA solicit the participation of other relevant federal agencies.* These include NIH, NSF, the Department of Energy, and the Department of Agriculture (USDA), among others.

Finally, considering the expense and commitment of resources to research in space biology and medicine, the committee recommended *that cooperation between international partners be explored to the fullest extent in undertaking the requisite research strategies required for a sustained human presence in space.* This approach is particularly relevant with respect to our foreign partners as design and development of Space Station Freedom proceed.

PROGRESS

A number of the recommendations outlined in the 1979, 1987, and 1988 reports cited above have been fully implemented. For example, NASA has undertaken a reorganization of its advisory structure in the area of life sciences and has created several discipline working groups along the subdiscipline lines recommended. NRAs and AOs for forthcoming missions have become more highly focused. NASA has signed a Memorandum of Understanding (MOU) with NIH regarding research efforts in space biology and medicine and has initiated contact with NSF and the USDA. The level of cooperation with foreign partners has increased dramatically and includes not only joint efforts concerning utilization of Space Station Freedom but also joint missions on forthcoming Shuttle flights and possibly even sharing of data, facilities, and equipment, and flying of experiments on COSMOS with the Soviet Union. Finally, NASA is in the process of establishing Specialized Centers of Research and Training (NSCORTs) to increase the scope of its interaction with the academic community. The initial NSCORTs, to be established beginning in 1991, will be focused on gravitational biology, environmental health, and bioregenerative life support. This is an excellent approach for increasing the science community's involvement in space biology and medicine and will focus research efforts in areas critical to a long-term human presence in space. In summary, NASA, and specifically life sciences within NASA, has made a significant and noteworthy effort to implement many of the recommendations contained in CSBM reports.

LACK OF PROGRESS

There has been essentially no progress in addressing recommendations associated with the design and utilization of the presently planned space station. At this time, there is no evidence to suggest that the recommendation for a dedicated life science laboratory has been seriously considered. Current plans call for sharing the U.S. and other lab modules between two major users, biology/medicine and microgravity (materials science) research. If this decision is final, it could significantly decrease the scientific return from Space Station utilization for both disciplines because of differing laboratory environment requirements.

The recommendation that 3- to 6-month blocks of time on Space Station be assigned to specific subdisciplines of space biology and medicine is not included in current planning. Again, the potential for meaningful scientific return will be seriously diminished if current plans proceed.

Finally, while virtually every internal and external life sciences advisory group over the last 20 years has emphasized the critical need for a VFC in space, this facility continues to be the subject of debate. In fairness, the debate has not been so much at the level of NASA as in the Congress.

There appears to be a major asynchrony between the research time that will be available on the station to solve the concerns associated with human adaptation to a microgravity environment and national goals and timetables associated with human exploration.

In terms of the broad issues associated with data management, it is imperative that the "user community" be involved in the design and utilization of the appropriate hardware and software. It has become clear that NASA does not have a central data base with all results from previous flights relevant to space biology and medicine. Similarly, a central data base does not exist on the results obtained from ground-based experiments relevant to the various subdisciplines. Individual investigators currently have no way to search archives for the literature on space biology and medicine. This is especially serious since much of the data which do exist have not been published in the open literature; they exist solely in NASA technical bulletins and the like. This practice opens the possibility that, among other things, the design of new experiments can become needlessly repetitive. There is some indication that NASA is aware of this critical deficiency and is planning steps to resolve the problem. We believe it essential to do so.

3

Human Physiology

Physiology is the science of normal vital processes of animal organisms. It is a fundamental science in its own right, but it is also the bridge to the practical problems of human health and performance in space. Problems produced by disordered physiology in space include a steady loss of bone and muscle, the cause of which has yet to be determined; abnormalities of the cardiovascular system related to pooling of blood in the chest and head; and motion sickness and the difficulties it causes not only in terms of astronaut discomfort but also in their performance in handling spacecraft. The committee has had, over the years, access to anecdotal data from the Soviet space program. This anecdotal information is, while interesting, not sufficiently reliable for drawing conclusions or in planning the U.S. program for a number of reasons. There are differences in experimental protocols and controls in laboratory equipment, and the Soviets do not publish their results in refereed scientific journals. However, increased recent cooperative activities between the Soviets and the United States suggest promise for the future in standardized experimental procedures and data exchange. In some areas of physiology, there has been appreciable progress.

Much remains to be done before we can be certain that humans can stay healthy and perform well for extended periods in space. The physiological problems of concern are divided into three main topics in order of priority regarding their importance to extended human space travel. First and of greatest concern is bone, muscle, and mineral metabolism; second, cardiovascular and homeostatic functions; and third, sensorimotor integration.

BONE, MUSCLE, AND MINERAL METABOLISM

Status of the Discipline

The bone and muscle atrophy that occurs in the microgravity environment is a severe hurdle to an extended human presence in space. Although astronauts are subject to elevated urinary calcium and increased risk of kidney stones, the most significant risk to the musculoskeletal system may only be realized on return to normal gravitational fields. As emphasized in both the 1987 and 1988 CSBM reports, very little regarding the etiology of space-induced osteopenia (bone loss) or muscle atrophy is understood. While countermeasures to inhibit or prevent this bone loss are imperative, this goal will only be realized following an improved understanding of the basic cellular mechanisms responsible for the maintenance of muscle and bone mass. As importantly, the interaction of microgravity with various risk factors (e.g., age, gender, race, nutrition) must be established to ensure that the musculoskeletal integrity of payload specialists of diverse backgrounds will not be compromised by extended spaceflight.

Major Goals

As recommended in the Goldberg Strategy, eight major goals have been defined for the science of bone and mineral metabolism. In brief, they are as follows: (1) determine the temporal sequence of bone remodeling in response to microgravity; (2) establish the reversibility of this process on return to 1 g; (3) establish the relationship between muscle activity and bone function; (4) investigate countermeasures to prevent bone loss; (5) study the cellular mechanisms responsible for bone loss; (6) study the interdependence of calcium homeostasis and bone remodeling; (7) determine the etiology of pathologic calcification; and (8) establish the biomechanics of the skeleton under microgravity conditions. Given these goals, it is clear that it will be difficult to separate operational considerations (astronaut health, safety, and performance) from the basic scientific questions.

In light of these goals, it must be emphasized that understanding the etiology of osteopenia is the focus of an enormous research program within the National Institutes of Health. The extent of NIH programs, which include Centers of Excellence (COE) and Specialized Centers of Research (SCOR), underscores our nation's commitment to improving an understanding of these disease processes. Multidisciplinary studies address issues such as promotion of bone formation, inhibition of resorption, biomechanics of skeletal modeling and remodeling. Also under development are treatment modalities such as diphosphonates, fluorides, estrogen, calcitonin, calcium, electrical and mechanical intervention, and exercise, with the aim of inhibiting or reversing bone destruction. These studies range from the structural organization of the bone mineral to human clinical trials of these novel treatment prophylaxes, and have had a great impact on the way we perceive the musculoskeletal problems that parallel aging, menopause, hyperparathyroidism, diabetes, poor nutrition, immobilization, and extended bedrest. Every effort should be made to ensure that the goal-oriented NASA mission concerning musculoskeletal science exploits the knowledge base derived from the scientific advances of other agencies.

Progress

Significant progress has been achieved by NASA scientists over the past five years. The initial mandates of the Goldberg committee have been addressed, and the animal models developed by NASA personnel and their colleagues have provided some insight into the skeletal tissue response to decreased functional stimuli. Whole animal studies have been performed that demonstrate the compromising effects of microgravity, and some information has been generated regarding the temporal sequence of tissue response to disuse. Additionally, human studies correlating inactivity (bedrest) to factors such as diminished bone mass and increased urinary calcium have proven useful models to understanding the risks of skeletal fracture or mineral disorders during extended spaceflight.

Lack of Progress

While significant work has been done, and NASA investigators have established collaborations with superb musculoskeletal and connective tissue scientists at a number of university centers, it must also be emphasized that the directions for future programs are not yet clear. Aside from the general hypothesis that gravity serves as a signal to the regulation of bone mass, it is not clear if NASA investigators have constructed a series of basic testable hypotheses regarding the cellular mechanisms that regulate this response. Nor is it clear whether the models currently used will be appropriate to deal with the next phase of study, i.e., countermeasures to prevent bone loss. Of the eight major scientific goals of the Goldberg Strategy, only the first has been addressed. Unfortunately, even this information is limited by the dissimilarities between the animal model chosen and normal human physiology. Considering the current focus on singular methodologies, it is unclear how NASA will investigate the remaining seven research objectives.

Specifically, it is not clear if the principal animal model of NASA scientists, the rat tail suspension model of osteopenia, has demonstrated either a repeatable bone loss or how this skeletal response correlates to that experienced in microgravity. Nor is it clear how this model will be developed to address issues such as a minimum strain environment capable of retaining bone mass, or the correlation of sites of bone formation and resorption to specific aspects of the bone's mechanical environment (i.e., what are the changes in the lower limbs' stress/strain environment following suspension?). Complementary analytical/experimental models should be developed immediately which can test (and validate) proposed hypotheses of bone remodeling. In addition, the limitations of the rat as a model, which responds to microgravity by inhibited growth, should be contrasted with the human condition of accelerated resorption. It must be determined whether commitment to a singular animal model is an efficient and effective means of addressing the range of scientific goals within the Goldberg Strategy.

Simultaneously, a realistic and reasonable time frame for addressing the scientific issues of the Goldberg Strategy must be implemented. If certain scientific criteria are not met within these temporal constraints, new avenues should be established immediately; the risk of overcommitting to a single model must be diminished. Indeed, despite the extreme efforts by NASA investigators regarding their models for osteopenia, a great reluctance remains within the general scientific community to accept the extrapolations to the human condition. Importantly, the research program must move to *beyond* the phenomenologic (i.e., bone tissue response to microgravity) to issues of mechanism. A commitment to in vitro studies of bone cells, to study mechanisms of perception and response to physical factors (e.g., mechanical/electrical), should be implemented immediately. Additional alternative strategies and technologies (i.e., molecular biology and structural biology) within the musculoskeletal system should also be considered. Before reasonable prophylaxes for bone loss can be developed, the basic cell and subcellular mechanisms responsible for this osteopenia must be understood. Finally, it is clear that this work will provide important spinoff technology for the treatment of earth-based musculoskeletal disorders.

CARDIOVASCULAR AND OTHER HOMEOSTATIC SYSTEMS

This section describes the problems, goals, and recommendations for space-related and in-flight studies of four broad areas of adaptive and protective physiology. The cardiovascular and neuroendocrine elements of the circulatory system are addressed in two areas, focusing respectively on basic cardiovascular functions and the influences of regulatory systems on these functions. The third area includes immunology, hematopoiesis, and wound healing, which share cellular constituents and the recognition of regulatory proteins. The fourth area, nutrition and gastroentology, describes the lack of major problems and discusses the absence of a relationship to serious problems with other systems.

Circulatory Adjustments

Cardiovascular physiology is a high-priority area for NASA. To a large extent, gravitational forces determine the distribution of intravascular and intracardiac pressures and volumes. The cardiovascular system appears to function normally during short-term exposure to microgravity. However, clinically significant dysfunction is often apparent during readaptation to 1 g and is likely amplified with prolonged spaceflight. In addition, prolonged exposure to the altered loading conditions of microgravity is considered to be a potential cause of irreversible functional and structural changes.

Status of the Discipline

Many important features of cardiovascular adaptation in microgravity and re-adaptation at 1 g have been identified during the past 20 years. There is an early shift of body fluids toward the head, followed by a loss of intravascular and interstitial fluid volume with decreased heart size. A fall in blood pressure on changing from a supine to an erect posture with consequent light headedness or fainting (orthostatic intolerance) and decreased exercise capacity are usually evident after return to Earth. Simple descriptive observations have been made in many subjects, but our understanding of the mechanisms responsible for adaptation to microgravity and readaptation to 1 g is incomplete.

Major Goals

The 1988 Life Sciences Task Group report states very clearly that biological and medical research in space should include different species and be performed at all levels of integration. This approach requires the development of a set of animal holding facilities, experiment modules, and work stations for Shuttle use and a wide range of facilities for the Space Station, including a large variable force centrifuge (VFC). This same report also makes the point that all crew members should contribute to a life sciences/space medicine data base by serving as subjects in biomedical studies.

The report defines several important cardiovascular research areas including (1) cardiovascular and systemic responses to the initial fluid shift and their interactions; (2) mechanisms responsible for postflight orthostatic intolerance; (3) the long-term, possibly irreversible, effects of the altered loading conditions in microgravity, e.g., atrophy of heart muscle, chronic orthostatic hypotension; and (4) maintenance of reflex control of blood pressure.

Specific high-priority areas of investigation include (1) the role of exercise and physical fitness before, during, and after flight; (2) countermeasures against cardiovascular abnormal function and rehabilitation after long flights; (3) validation of ground-based models of microgravity for short-term and long-term studies, and (4) characterization of drug actions and metabolism in microgravity.

The Goldberg Strategy also defines a series of scientific goals: (1) to understand acute (0 to 2 weeks), medium-term (2 weeks to 3 months), and long-term (3 months to several years) cardiovascular and pulmonary adaptation to microgravity; (2) to examine the validity of ground-based models of microgravity and to determine whether actual microgravity offers any unique advantages for cardiovascular studies; and (3) to define measures that will hasten human adaptation to microgravity and return to a 1-g environment.

Progress

Studies performed by Johnson Space Center investigators before, during, and after several Shuttle flights have provided echocardiographic data on cardiac dimensions and function and also explored the use of oral saline loading as a countermeasure against postflight orthostatic hypotension. Studies in progress deal with reflex regulation of blood pressure (baroreceptor reflex).

A series of more detailed and comprehensive cardiovascular and pulmonary experiments will occur on future Spacelab flights to meet many of the objectives listed in the Goldberg Strategy for short-term flights, ranging from direct measurement of central venous pressure and cardiac output during maximal and minimal exercise, characterization of the baroreceptor reflex, and detailed study of the effects of microgravity on body fluid volumes and their regulation.

The validity of head-down tilt as a method of simulation of microgravity will also be explored. SLS-1 will also initiate the use of animals (rats) for the study of cardiovascular physiology. The SLS-1 experiments are also an important component in a unique NASA pilot program in which experiments planned for a Spacelab life sciences (SLS) flight will be discussed as a means of enhancing elementary and secondary school science classes.

With joint participation by NASA, ESA, and the German Space Program, the D-2 flight will extend the range of measurements and interventions with detailed observations during intravenous fluid load and lower body negative pressure.

Lack of Progress

The number of subjects available for physiological and medical studies is generally fewer than would be desired. The Life Sciences Task Group of the Space Science of the Twenty-First Century study noted that increased participation by the crew in biomedical studies during all flights is an important objective.

A better understanding of cardiovascular and pulmonary physiology at all levels of investigation is still a major science goal and a requirement for a significant rational approach to space medicine, including the development of scientifically sound countermeasures against, and adaptation to, changing gravitational forces.

There are only limited Soviet data on the cardiovascular consequences of *prolonged* exposure to microgravity. The U.S. data are limited to the Skylab exposure, which produced data on medium-term exposure. Much more information is needed on cardiovascular and pulmonary responses to the changing loading conditions in space, and on the interactions between the cardiovascular system and its control mechanisms.

Potential interactions between the space environment and cardiovascular and pulmonary physiology modified by disease processes or pharmacological agents have not been addressed and also need attention.

Hormones and Stress

Hormones that affect the cardiovascular system are of great importance to NASA and should be considered in the context of the cardiovascular changes that occur in space cardiovascular deconditioning.

Progress

Recommended measurements of many hormones have already been made in astronauts who made the 28-, 59-, and 84-day flights in Skylab in 1973 and 1974. Two comments are relevant in terms of these observations. First, it would be beneficial to repeat them to increase the number of subjects studied in order to get a better idea of biological variations. Second, it would be wise to repeat them using the more refined methods that have been developed since 1974. A controlled study of blood and urinary levels of the relevant hormones is planned for SLS-1. Experiments are planned for mission specialists who will be concerned with laboratory experiments and will not be involved in flying the spacecraft. Consequently, better data should be obtained.

Lack of Progress

The observations planned to be made on SLS-1 are still far from ideal. Even if the best data are obtained, it remains only descriptive. It is important that ongoing ground-based investigations of the fundamental mechanisms involved in producing the hormonal responses observed in space be carried out in humans and animals. One concern not specifically addressed in either the 1979 or the 1987 report is the hormonal response to stress. For instance, the adrenal hormone, cortisol, was measured in Skylab and was found to be elevated from time to time during spaceflight. It would be appropriate to carry out a more detailed, better controlled study of the pituitary hormone ACTH (that stimulates the release of cortisol from the adrenal cortex) and cortisol secretion on a future SLS mission and on the Space Station. Measurements of ACTH and cortisol are also relevant to investigation of circadian rhythms in space. This topic is covered in detail in another part of this report (see Chapter 4 discussion regarding circadian rhythms).

Many experiments that have examined physiological and behavioral variables in space have failed to take into account the rhythmic nature of these factors. For example, it is not possible to accurately determine the overall pattern of adrenal cortisol secretion (a stress-related hormone) by simply taking a single blood sample on a daily basis. However, NASA has carried out such studies with the result that contradictory findings have been obtained within and between different missions. All experimental protocols that propose to use single point measurements for variables that show pronounced pulsatile and/or circadian variations need to be reexamined to insure that the data, which will be collected at enormous costs, are valid and will not be meaningless to the scientific community.

Immunology, Hematopoiesis, and Wound Healing

Status of Discipline

Immune cells in mammalian bone marrow and lymphoid organs, such as the thymus and spleen, provide integrated and adaptive host defense against infections, other external environmental challenges, and deviations in host cellular growth and differentiation. Many of the cells and proteins of the immune system, which initiate and regulate lymphocyte and antibody responses, also control the production and functions of cells in the blood and connective tissues.

Major Problems and Goals

The importance of investigations of immunology and related systems in space was suggested initially by the findings of abnormalities in human lymphocytes, red blood cells, and other blood cells on Spacelab Mission D-1, and rat immunity on unmanned Soviet spaceflights, as well as by a few anecdotal reports of an increased incidence of cutaneous, gastrointestinal, and renal infections in humans on Russian and U.S. spaceflights.

Progress

Most studies of immunity in space have been directed to the detection of abnormalities in human and animal lymphocyte numbers and morphology, their proliferation and synthesis of immunological proteins in response to bacterial products, and the serum concentrations of gamma globulins important in immunity. Other human white blood cells prepared prior to launch also showed impaired function after spaceflight, relative to ground controls, when studied after landing.

Spaceflight results in significant reductions in both plasma volume and red blood cell mass within days, and reversal of both after a few weeks. Although some reports have cited minor abnormalities in red blood cell biochemistry, no major primary metabolic or structural defects have been documented. There are plans to measure aspects of iron uptake and storage in bone marrow.

Lack of Progress

The very uncommon occurrence of serious infections during spaceflight, despite the apparently profound laboratory defects in some immune functions *in vitro*, casts some doubt on the preliminary findings of alterations of immune responses. However, the grave consequences of any attenuation of adaptive host defense during spaceflight and the important roles of immune proteins as regulators of non-immune blood cells and connective tissue cell generation and functions, such as wound healing, mandate recommendations for more investigations of immune system functions. Our current understanding of immunity in space is not commensurate with the goals proposed in *Space Science in the Twenty-First Century—Imperatives for the Decades 1995-2015, Life Sciences* and earlier sets of recommendations.

Above all else in practical importance are *in vivo* studies designed to elucidate any abnormalities in integrated immune, hematopoietic, and tissue-healing responses of humans. The critical need for 1-g in-flight controls cannot be overemphasized, if the resultant data are to be analyzed rigorously for specific mechanisms and to provide unequivocal clinical guidelines for countering any abnormalities. The availability now of stimuli and inhibitors of immune and bone marrow cell functions, derived from genetic techniques, provides an opportunity to detect and possibly correct any excesses or deficiencies of activity or any regulatory abnormalities. Studies of the mechanisms for reduction in red blood cell mass with spaceflight and of effective countermeasures for attenuated compensation to bleeding and blood clot formation are recommended.

Gastrointestinal System and Nutrition

Major Goals

A Strategy for Space Biology and Medical Science for the 1980s and 1990s recommends studies of caloric needs, nitrogen balance, supplements to the basic diet during spaceflight, and the energetic requirements of work in space. *Space Science in the Twenty-First Century— Life Sciences* discusses nutritional requirements in the context of integrated functions and closed ecological life support systems (CELSS).

Progress

Appreciable data are available on energy expenditures and motor performance. In experiments carried out to date, nitrogen balance in space has been negative, probably because dietary intake was often inadequate in view of space motion sickness and crowded schedules and because of muscle wasting. The amounts of vitamins and minerals provided to the astronauts more than meet the *Recommended Dietary Allowances*, (Tenth edition, National Academy Press, Washington, D.C., 1991), and there is no evidence indicating any additional special needs in space. Little is known of possible minor changes in bowel, liver, pancreatic, and salivary functions during spaceflight, although there are plans to study gastric emptying and gastrointestinal motility and their relationship to movement of body fluids.

Lack of Progress

Very few significant problems have been encountered in space with nutrition or the gastrointestinal system to date. However, NASA has plans to conduct investigations on areas that may be potential problems on future Shuttle flights.

SENSORIMOTOR INTEGRATION

Status of Discipline

The changes in the gravito-inertial environment that inevitably occur during a space mission lead to disturbances of sensorimotor function including impaired spatial orientation, instability of posture and gaze, and motion sickness. Fortunately the central nervous system (CNS) adapts to those environmental changes within a few days so that the problems are of relatively short duration *provided a constant environment is maintained.* There are two caveats to this assessment of relative risk. One is that gravito-inertial changes occur at the most critical parts of a mission—during takeoff or landing. For instance, sensorimotor problems could impair crew effectiveness for several days after landing on Mars. The second caveat is that the use of a spinning spacecraft to eliminate problems of prolonged microgravity would entail repeated changes in gravito-inertial environment when it was necessary to change spin rates or to service different parts of the craft. To deal with these concerns, *A Strategy for Space Biology and Medical Science for the 1980s and 1990s* recommends experimentation in five areas: spatial orientation, postural mechanisms, vestibuloocular reflex (VOR), neural processing in the vestibular system, and motion sickness. Both in-flight and ground-based experiments were proposed in each area. Because the otolith organs of the vestibular labyrinth are the primary sensors informing the brain of the linear accelerations induced by gravity, the research recommended focuses on neural mechanisms that transform vestibular sensory inputs (especially those from otolith organs) into sensations, movements, and changes in homeostatic state (i.e., space adaptation sickness).

Major Goals

As indicated in the Goldberg Strategy, the neuronal mechanisms underlying a sense of *spatial orientation* are complex, as yet relatively poorly understood, and directly relevant to assuring effective functioning of humans involved in space missions. Accordingly, the strategy recommends a vigorous program of both ground-based and flight research aimed at better understanding these mechanisms as they operate on Earth, in space, and on return from low-gravity to high-gravity environments.

The report also recommends that the adaptation of the *posture control* system to microgravity be investigated by studying its input/output behavior before, during, and after flight.

The VOR, which converts head motion into compensatory eye movement, is both well understood and important in maintaining stable vision. However, it is not known how this reflex is affected by microgravity. Accordingly, the 1987 strategy report recommends that it be studied in microgravity.

The Goldberg Strategy concludes that *neural processing* mechanisms in the vestibular system must be investigated both in normal gravity and microgravity if NASA is to achieve its goal of controlling for sensorimotor changes and space adaptation sickness. These studies would focus on the vestibular nuclei of the brainstem, which serve both as the point at which vestibular signals from receptors of the vestibular labyrinth in the inner ear enter the brain and as a primary processing center for all signals—vestibular, visual, and proprioceptive (sense of body position)—that relate to orientation in space and to motor responses that maintain a stable orientation. In addition to their potential for solving operational problems, studies of vestibular nuclei in microgravity provide unique opportunities for understanding basic vestibular function since microgravity is the only environment in which certain aspects of semicircular canal and otolith function can be studied in isolation.

The report further concludes that one focus of sensorimotor studies must be upon the *adaptive mechanisms* that alter vestibular processing in response to altered feedback from the environment. Present knowledge indicates that it is these mechanisms that both allow an astronaut to adapt to microgravity conditions and then raise problems during return to normal gravity when the adapted responses are no longer appropriate. Further, space adaptation sickness may be a side effect of the adaptive process.

The Goldberg Strategy reviews uncertainties about the *etiology of motion sickness* and stresses the importance of vigorous research on this subject, with the shorter term objective of mitigating associated operational difficulties. Longer-term objectives include eliminating such difficulties as well as enhancing understanding of the function of the nervous system.

Progress

Overall, NASA has made a concerted effort to stimulate appropriate, quality research in the sensorimotor area. Space adaptation sickness (SAS) has clearly been perceived as an important problem that can only be solved by a combination of basic and more applied research. In the area of basic research, NASA has supported an array of external studies, which have included contributions by a number of leading vestibular physiologists. These have generated useful information in the areas of spatial sensation, vestibular processing, posture and gaze control, and vestibular-autonomic interactions. A promising development is the establishment of the Vestibular Research Facility (VRF) at Ames Research Center (ARC), which is becoming a focal point for state-of-the-art studies of the vestibular system in animals, involving investigators from many universities and other research institutions. If this facility can become a national resource and a site of projects with joint funding from NASA and NIH, its impact will be even greater. Whereas in some cases microgravity is a singular point so that behavior on-orbit cannot be extrapolated from measurements at various g levels on the ground, it is likely that some of the vestibular changes in going from 1 g to microgravity could be extrapolated from changes between a range of gravitational forces. Therefore, parallel development of the ARC centrifuge or other similar facilities as centers for vestibular studies could contribute to solving some of the outstanding questions.

Given the limited number of flight opportunities, considerable progress has also been made in studying human sensorimotor performance in microgravity. A great deal of this progress has resulted from the well-planned experiments conducted by groups at MIT as well as in Canada and Europe. Tantalizing preliminary results and interesting new concepts have resulted from their studies on the Spacelab and D-1 missions. Follow-up studies on SLS-1, IML-1 and D-2 missions should be very productive, but it must be emphasized that extensive replications will be needed before any of these preliminary results can be accepted with confidence.

As set forth in the Goldberg Strategy, what is needed now in vestibuloocular studies are 3-dimensional analysis and modeling of the VOR to account for changes in microgravity and investigation of the effects of conflict between visual, vestibular, and proprioceptive stimuli on gaze stability and on the adaptive systems that regulate the VOR. Studies of vestibuloocular and optokinetic (visually induced) reflexes scheduled for SLS-1 will address these issues as will planned ground-based studies of adaptive changes in otolith-ocular reflex. Thus work in this area is on track.

Progress in meeting the objectives set forth in the 1987 report regarding neural processing can be summarized as follows:

Studies of *structural changes in labyrinthine receptors* in microgravity are under way and will be continued on SLS-1. In the future, it will be necessary to perform control experiments in which some animals are maintained at 1 g on orbit using the centrifuge so that we can be confident that any changes observed are due to exposure to microgravity rather than other factors.

Studies of *vestibular afferent signals* (i.e., those nerves that transmit impulses to the CNS) in microgravity and studies of the role of gravity in neonatal development of sensorimotor functions will require periods of a month or more in orbit. Plans should be formulated to perform these studies on Space Station Freedom and prerequisite ground-based development should be initiated. The CSBM is not aware of any such activities at the present time.

The Goldberg Strategy recommends "a vigorous, ground-based program of research involving the use of modern physiological methods" to investigate the *central processing* of vestibular sensory information, especially that from the gravity-sensitive otolith organs. NASA is to be commended for establishing the VRF Program and a good series of university-based studies to address this goal. Further emphasis on single neuron recordings will be needed to attain the goal of understanding those elements of vestibular processing that are related to sensorimotor changes in microgravity. Even more important is the development of *neural models* of the vestibular system that could both generate hypotheses for experimental testing and assist in the interpretation of experimental results. If properly managed, the nascent modeling effort at the ARC could provide a national resource for such theoretical studies of vestibular systems.

As appropriate, NASA appears to have been pursuing an active multidimensional research program in understanding *motion sickness*.

Lack of Progress

Within this generally positive context, NASA might consider some fine tuning of its program related to spatial orientation along the following lines. The Goldberg report stresses that vestibular signals are only one of the inputs influencing spatial orientation, with other kinds of inputs certainly involved and perhaps determinative under particular sets of circumstances. The issue of whether observed instances of sensorimotor difficulties are, or are not, entirely a function of disturbed vestibular processing has become of operational importance in connection with ongoing and projected increases in mission duration (letter to Administrator Truly, NASA, regarding extended duration orbiter medical program, December 20, 1989). In this connection, both increased attention to possible nonvestibular influences on spatial orientation, and improved monitoring of astronaut performance (as recommended in that letter), would seem desirable. It also appears that these and other observed disturbances may relate not so much to steady-state conditions as to fluctuating environments, such as the changing gravity loads associated with return from orbit. This suggests that in the ground-based program increased attention might profitably be given to the program of spatial orientation in fluctuating environments. Finally, the breadth of the problem of spatial orientation is such that efforts in this area will probably be most productive when considered in connection with those in possibly related areas, including, for example, not only other sensorimotor subareas but also chronobiology and human behavior generally. NASA might want to explore ways to assure effective interactions along these lines, possibly by way of organizing a meeting focused on the spatial orientation problem but involving investigators from a number of different relevant areas.

NASA should consider increasing its support of ground-based studies of postural mechanisms to complement the flight studies since significant changes in eye-hand coordination and in postural responses to applied perturbations are being found. When a sufficient level of understanding has been reached, it may become possible to train astronauts to adopt postural strategies that will be useful in altered gravity environments as suggested by the Goldberg Strategy.

The problem of VOR has already been discussed and NASA appears to be headed in the correct direction. Similarly, issues related to neural processing are being dealt with properly, although a great deal of work remains to be done.

It remains the case that no single hypothesis as to the origin of space motion sickness fits all existing observations, and no single therapeutic regime with assured effectiveness has yet been developed. While existing lines of research may yet achieve one or both of these objectives, it may be that the time has come to take a fresh look at whether the syndrome, with individual variations in the expression and intensity of components, is actually several distinguishable syndromes. A recognition of this fact could accelerate both therapeutic and basic research. When viewed in the light of resources required to develop new therapeutic drugs, the amount of funding for this key area is clearly inadequate. If NASA requires a solution to this problem, ways need to be found to mobilize a much larger joint NASA/university initiative in the area.

4

Behavior, Performance, and Human Factors

All the major reports over the past decade on research needs for life sciences in support of human spaceflight—*Life Beyond the Earth's Environment, The Biology of Living Organisms in Space*; *A Strategy for Space Biology and Medical Science for the 1980s and 1990s*; *Leadership and America's Future in Space* (NASA, A Report to the Administrator by Sally K. Ride, August 1987); *Exploring the Living Universe—A Strategy for Space Life Sciences* (NASA, Washington, D.C., June 1988), and *Space Science in the Twenty-First Century—Imperatives for a New Decade—Overview* and *Life Sciences*—recognize problems in human behavior during long-duration missions such as Freedom, a manned lunar base, and a Martian outpost. It is not known whether missions longer than one year are endurable with present environments and organizations. Additional knowledge may be critical to enable such missions. Critical issues are the selection and training of candidates for spaceflight, the composition of crews, the physical characteristics of the spacecraft, and the organization and structure of missions including the requirements for successful leadership and how to employ automation effectively while providing meaningful work and optimizing performance and satisfaction in work. Social issues may be critical, limiting factors in the exploration of space. Small groups of individuals have to operate effectively and harmoniously for prolonged periods, separated from families and customary sources of support. Circadian (i.e., 24-hr) rhythms in the space environment need to be investigated to ensure maximum health, productivity, and performance.

STATUS OF DISCIPLINES

Knowledge of the psychological principles needed to optimize the behavioral effectiveness of individuals in groups in demanding situations is primitive. The recognition that such knowledge can be acquired by a systematic, experimental approach is recent. Findings have not had time to have major impact on the design and execution of space programs.

Organizational influences on flight crew performance and attitudes are now being investigated by NASA's Ames Research Center (ARC). Training strategies to improve the quality of group decision making, information transfer, and leadership are also under study in civil and military aviation. These studies provide a starting point for comparable research in spaceflight.

Most research to date has studied individuals and groups over relatively short periods of time. Longitudinal studies of individuals and groups over extended periods are needed.

Circadian rhythms continue in space but rhythm disturbances in plants and animals, including humans, have been noted. Future studies are required to determine whether the changes have adverse effects on crew performance and health, and to determine the most optimal environmental conditions for maintaining synchronization of circadian rhythms.

MAJOR SCIENTIFIC GOALS

The major goals are to optimize the efficiency, safety, and satisfaction of crews in long-term spaceflight by discovering how to optimize (1) the environment of spacecraft, (2) human interfaces with equipment, (3) work and leisure schedules, (4) social organization, and (5) selection and training of crews.

PROGRESS

Research employing appropriate methods that was recommended in the Goldberg Strategy is now in progress. However, progress has been slow, commensurate with the minimal resources provided, especially on group and organizational factors.

As mandated by the Aviation Safety Research Act of 1988, behavioral science research in aviation is being supported by the Federal Aviation Administration (FAA) as well as NASA. Congressional support for research is likely to continue in the aviation area. Although useful, such studies cannot provide all of the information needed for long-duration missions. Neither the NSF nor the National Institute of Mental Health (NIMH) will support the major part of the research needed in this area.

Opportunities for experimentation in actual spaceflight will always be extremely limited. Therefore, a sound basis must be developed in ground-based studies in a variety of research settings. It is self-evident that assessment of factors that will maintain efficiency and satisfactions in small groups confined for 3 or 4 years will necessitate studies on groups confined for 3 or 4 years. Such studies should be completed on Earth before missions of such duration are conducted, or even planned, in space so that information can guide designs of environment and life schedules. It is disappointing that there do not seem to be plans in NASA to even start working toward such long-term confinements. The results from analog settings of both laboratory and field experiments must then be validated in operational environments.

Environmental Factors

Modest investigations continue on influences of features of spacecraft design on performance and adjustment. Research into the impact of automation on operator performance and reactions is supported in the Aerospace Human Factors Division at NASA-ARC. The object is to determine the most effective combinations of automation and control by human operators.

Individual Factors

Investigations into ability, personality, and motivational factors relevant to crew selection are under way at ARC in aviation and at Johnson Space Center in space research. A common core of personality attributes is under investigation. These factors have been validated as critical to air transport crew effectiveness. In space-related investigations, an experimental battery of tests was administered to applicants for the most recent class of astronauts and is being contrasted with results obtained using clinical measures. The goal is to move beyond the process of screening out individuals exhibiting significant psychopathology toward selection based on a psychological profile associated with excellence in performance and adjustment. Behavioral criteria for selection are under development and will be used within the astronaut corps to validate the constructs under consideration. This work follows recommendations made in the Goldberg Strategy and is the first attempt to develop more precise methods of selection for spaceflight.

Many studies have been carried out to determine the optimum work schedules for maximizing human performance under normal conditions on Earth. Aviation research is examining crew fatigue and scheduling issues in both short-haul and transmeridian flights. Ames is sponsoring research into cognitive functions relevant to crew performance as part of its aviation human factors program. The results should have applications in spaceflight, but much further research and validation are necessary. In particular, it is crucial to determine whether task performance changes when individuals are fatigued and/or face various stressors.

Group and Organizational Factors

A study of the organization and communications practices among teams working at Kennedy Space Center on the integration of payloads for orbiters is being conducted by ARC. Work is beginning on multinational team behavior at Johnson Space Center. With the exception of these studies and work noted in aviation, the topic remains unexplored.

Circadian Rhythms

Studies on the D-1 mission and STS-9 clearly demonstrated that circadian rhythms persist in space but show abnormal amplitudes and phase relationships to the light-dark cycle. Such abnormalities have been observed in plants, rats, and monkeys. "Sleep and Circadian Rhythms" was included in the NASA Announcement of Opportunity for IML-2. In addition, NASA sponsored a July 1990 workshop on circadian rhythms and space.

LACK OF PROGRESS

Historically, the lack of progress in space-related behavioral research has been due to a lack of research funding and to the limited access to astronauts for such studies. The importance of psychological factors for effective long-duration spaceflight is increasingly accepted, but financial support continues to be at a very low level.

Most research has explored only a single domain of behavior (e.g., effects of fatigue or personality on a particular performance). Communication and collaboration across subdisciplines to consider the multiple determinants of reactions need to be fostered to permit integration. Although NASA has sponsored some interchange among investigators (e.g., the Life Sciences in Space Symposium, June 1987), the effort has been insufficient to develop comprehensive approaches to research. In space and other operational settings, psychological factors are likely to interact in complex ways to determine individual and group reactions. Systematic studies in both laboratory and operational settings need to be increased. Issues include scheduling, authority structure, and provision of meaningful work activities for crews on long-duration missions.

One indicator that behavioral issues are not fully integrated into many research programs comes from funded research into potential problems associated with increasing the duration of STS flights (letter to Administrator Truly, NASA, regarding the extended duration orbiter medical program,

December 20, 1989). Questions raised about difficulties associated with increasing mission duration center on the degradation of crew ability to land the orbiter successfully and to egress unaided after longer periods in orbit. The proposed program of research concentrates on standard biomedical measures. While there is recognition of the fact that the critical outcome is *behavioral*, the only actions undertaken have been to start superficial task analyses of the landing and egress behaviors. There is a need to develop *behavioral* criteria that reflect these tasks and to relate performance in these areas to biomedical indices isolated in either flight or ground-based studies. The proposed Biomedical Monitoring and Countermeasures Project (BMAC) is an expansion and extension of the EDO medical program. This is a newly organized project that is aimed at developing research to optimize crew performance on Freedom. Critical factors for BMAC are mental and social well-being, normal body state, and normal risk levels. A lack of useful behavioral data from EDO has the potential of limiting the effectiveness of BMAC.

Studies on small groups confined in close quarters can be conducted at any convenient location and can mimic all environmental features of spaceflight except microgravity. A challenge is to develop conditions to enable subjects to remain effective and satisfied during confinement for increasingly long periods, up to as long as three or four years. It is likely that the provision of meaningful, indeed engrossing, work for a considerable part of each day will be of great importance. Development of conditions will be iterative and hence very time-consuming. Only when extensive studies have elucidated major factors maintaining performance and satisfaction for long periods in confinement will it be necessary to conduct studies in more expensive and inconvenient sites such as the seabed or Antarctica, which because they are more inescapable, have more prima facie validity. Chances to collect opportunistic data in isolated, natural settings should not, of course, be neglected. Final validation of systems for long-term spaceflight will come only when long voyages are actually made, just as proof of long-term survival in microgravity had to wait for prolonged orbital flights. The research on Earth will surely establish factors that enhance effectiveness, so justifying the considerable expense of ground-based studies.

Few attempts have been made to follow human circadian rhythms in space in any detailed fashion. While isolated nights of sleep have been recorded on three Spacelab missions, rhythms in the sleep-wake cycle have not been monitored in space, and very little is known about other circadian functions in the space environment, particularly the circadian patterns of various hormones and/or metabolic productivity.

5

Developmental and Cell Biology

The discipline of developmental biology concerns all aspects of the lifespan of an organism including gamete production, fertilization, embryogenesis, implantation (in mammals), organogenesis, and maturational changes that occur in tissues and organs after birth (postnatal development). Although development is a continuous process, research frequently emphasizes experimental questions concerned with specific developmental time periods (e.g., the period of organogenesis). The most pressing issue in developmental biology for space concerns the question, can any organism undergo a complete life cycle (fertilization through production of viable gametes in the adult) in the microgravity of space? Alternatively, are there developmental phases so sensitive to microgravity that normal development from fertilization of an egg to fertile adults does not occur in space?

It is difficult to study the entire developmental process in a single organism, partly because of variable length in the life cycle. More importantly, different kinds of organisms exhibit diverse strategies for undertaking the complex processes associated with differentiation and organogenesis. Some invertebrate animals display highly ordered patterns of cell division, allowing precise studies on cell lineage. Thus, if microgravity leads to developmental abnormalities that are evident in the adult, the time at which various cells were altered can be traced backward in time. The development of vertebrates presents aspects that cannot be evaluated in invertebrates. For these reasons, research strategies require the use of several different kinds of organisms to fully understand effects of the space environment on developmental processes. Very little information on the development of animals in space exists. Much of the initial research on animal development will be of necessity observational in nature. Major advances in developmental biology increasingly have relied on studies at the level of individual cells.

The field of cell biology was not discussed in the Goldberg Strategy, but recent evidence obtained from Shuttle flights indicates that such problems need to be addressed. Progress in cell biology research represents one section of this chapter.

Whereas the study of development in various organisms addresses many questions of basic interest, the study of mammals in particular has fundamental interest that relates not only to the health and welfare of humans in space but also to any long-term plans for establishment of human colonies in space. For these reasons, the chapter on developmental biology also includes a summary of progress concerned with human reproductive biology in space.

STATUS OF DISCIPLINE

Major advances have occurred in the field of developmental biology since preparation of the Goldberg Strategy, especially in the use of modern techniques of molecular biology to study developmental processes. The use of specific "molecular markers" has enabled investigators to more precisely define the temporal aspects of cell and tissue differentiation. Studies on the effects of changing environmental cues on gene expression during the process of differentiation have become readily feasible and are under intense investigation in numerous laboratories. Related to this, molecular studies concerned with the mechanism(s) by which cells communicate with each other in the establishment of patterns have opened up several new experimental approaches. However, these advances do not necessarily alter any of the recommendations delineated in the SSB reports referred to in this document. Those recommendations dealt with the necessity to acquire basic information on the ability of various organisms to undergo normal developmental processes in microgravity. Once the initial data base has been obtained, research strategies designed to evaluate the mechanism by which microgravity affects developmental processes need to include modern molecular approaches as well as the more classical techniques of developmental biology.

MAJOR GOALS

The Goldberg Strategy recommended that highest priority be given to experiments on model organisms to determine at the outset if development from fertilization through the formation of viable gametes in the next generation, i.e., egg to egg, can occur in the microgravity environment of space. In the event that normal development does not occur, the next recommended priority is to determine which period of development is most sensitive to microgravity. A second goal is to evaluate the use of the space environment as a tool to study specific developmental phenomena in ways that cannot be accomplished adequately in ground-based research.

To achieve these goals most efficiently, and within the limitations imposed by restricted flight opportunities, several model organisms were recommended for study in space including two representative invertebrate organisms, two representative nonmammalian vertebrates, and at least one representative mammal. Whereas it may not be necessary to stress all the specific organisms recommended in the Goldberg and earlier strategies, it is important to stress the recommendation to study the common house mouse as opposed to the more frequent emphasis by NASA on the use of rats. The house mouse is one-tenth the size of rats, has a relatively short gestation time, and has the advantage of an extensive genetic data base. These advantages allow studies in space revolving around the time of implantation and, on the same animals, the period of postnatal development. Finally, since the only rodent embryo that can be successfully cultured is the mouse embryo, it is possible to study phases of mammalian development in culture in parallel with studies on embryos *in vivo*. This would allow a distinction between maternal contributions to the embryo in a microgravity environment versus direct effects on the embryo.

PROGRESS

A number of diverse organisms have been subjected to microgravity for varying periods of time. These include invertebrates such as the fruit fly (Drosophila), vertebrates such as amphibians (Rana pipiens and Xenopus laevis), and mammals such as the rat. In addition, a number of eggs (fertilized prior to flight and fertilized in space) have been observed in microgravity including those of insects, frogs, and chickens. The results of these experiments have been inconsistent. Both normal and abnormal development has been observed, depending on the organism and the stage of development at which material was subjected to microgravity. However, in some cases, the length of the experiment was insufficient and almost none of the experiments have had an in-flight 1-g control. In addition, some of the "experiments" have been carried out under the auspices of undergraduate or high school students.

In spite of the above limitations, the available data imply that microgravity may have serious detrimental effects on certain developmental processes including oogenesis (the formation of eggs) and the development of certain organs, especially those involved in gravity sensing. The D-1 mission flown in 1985 was the first mission containing an on-board 1-g centrifuge. Results from experiments on that flight showed clear-cut effects of microgravity on certain phases in the development of two insect species. A disruption in oogenesis was observed as well as a reduction in the number of eggs laid and an effect of sex ratio among developing offspring. These experiments further documented an effect of radiation on developmental processes, including synergism between radiation and microgravity. At least two upcoming missions contain experiments concerned with the effects of microgravity on the development of selected organisms. These include the Spacelab-J mission, which contains an experiment on development of the embryonic axis in frogs. In addition, the IML-1 mission contains the "frog experiment," one to examine development of Drosophila, and one to examine development of an additional insect (*C. morosus*). Finally, the proposed Space Biology Initiative (SBI) plans four areas of study including gravitational biology. The latter references studies on the role of gravity in shaping the growth and development of individual cells, plants, and animals. However, as this initiative has not yet been approved as a formal new start, no specific experiments have yet been proposed or selected.

LACK OF PROGRESS

To our knowledge, no animal species has ever been carried through one complete life cycle in the microgravity of space. Virtually none of the model organisms referred to in the Goldberg Strategy have been flown. Although some of these organisms are included in forthcoming flights, it is less clear that the experiments are designed to answer the first priority question, development from egg to egg. No experiments dealing with postnatal development, a critical issue described in the Goldberg Strategy, have been conducted or are planned. In many instances these problems result from the fact that the most immediate forthcoming opportunities include experiments which have been in the pipeline for some time; the queue is long and arose prior to publication of the strategy. Perhaps the most informative mission, the D-1 flight, was planned almost entirely by the European Space Agency prior to release of the Goldberg Strategy. In other instances, planning has emphasized other experimental questions. For example, much of the research proposed for LifeSat (a free-flyer proposed for a new start in FY 1992) is concerned predominately with radiation effects and dosimetry. Prior to availability of a space station, LifeSat is the only opportunity for these investigations. LifeSat will allow for mission lengths of sufficient duration to accommodate one or more complete life cycles.

In short, the data base concerning the effects of microgravity on developmental processes remains almost nonexistent. Without question, the two major impediments to progress in developmental biology research have been, and remain, the low level of support for basic research in gravitational biology and extremely limited opportunities for access to space.

CELL BIOLOGY

Human lung cells were flown on Skylab 3 (1973) with the result that no significant changes were observed in cell function. This led to the view that microgravity has little effect at the level of individual cells. Research supported by the gravitational biology program has concentrated largely on whole organism studies. The 1987 strategy reflected this view and did not propose research to investigate changes at the cellular and molecular level. However, it has become clear, especially from D-1 flight data obtained after the publication of the Goldberg Strategy, that the microgravity environment of space does have effects at the cellular level.

Status of Discipline

As with developmental biology, major research advances have occurred in the field of cell biology, due in large part to use of molecular approaches to study cell function. For example, genes for many membrane receptors have now been cloned, allowing the design of experiments to determine how cells respond to specific extracellular signals. The intracellular pathways, which are activated (inactivated) in response to environmental cues, are being worked out in detail. Mechanisms by which cells synthesize and process proteins destined for secretion and proteins required in the construction of the cytoplasmic architecture (i.e., tubulin, actin, intermediate filament protein), are now amenable to precise analysis. Finally, several *in vitro* systems have been established in which the function of individual cells can be studied in detail. Of relevance to the goals of space biology and medicine, both bone and muscle cells can be induced to differentiate in culture. Thus, the opportunity exists to establish model systems using ground-based research that can then be used to study the effects of microgravity of specific cellular processes.

Progress

Changes in basic cellular and metabolic function were observed in tissues from animals on the Spacelab-3 mission (1985). These included reduced synthesis of RNA, reduced growth rates in blood cells, changes in tubulin and cytoskeleton synthesis and distribution, and changes in collagen secretion. The most complete studies carried out on the D-1 mission indicated significant alterations in single cells from both prokaryotes and eukaryotes. Thus, bacterial cells as well as cells of algae, slime mold, and protozoan organisms such as parameciums all grew more rapidly in space. Surprisingly, mammalian cells in tissue behaved in an opposite manner and divided more slowly in microgravity. The cause of this phenomenon is unknown, but the data suggest a reduced rate of glucose utilization and changes in membrane structure.

Lack of Progress

To our knowledge, no comprehensive experiments are planned to test the effects of microgravity on cell function, although additional studies are planned to confirm the microgravity effect on division of mammalian cells in culture. This is not, however, appropriately considered a lack of progress since the effects of microgravity on cell function have become apparent only recently. There is a need for the development of a comprehensive strategy in cell biology. Considering the long lead times for actual flight experiments, such strategies should become a high priority in order to allow for the conduct of appropriate experiments in the foreseeable future.

HUMAN REPRODUCTIVE BIOLOGY

Status of Discipline

Subsequent to publication of the Goldberg Strategy, there have been significant advances in reproductive technology which would enhance the ability to monitor men and women during spaceflight for their reproductive status. These include assays for the sex hormones that can be performed on urine and saliva samples, reducing the need for serum sampling of the crew. In addition, analysis of a number of key parameters of sperm function has become fully automated and computerized. These advances enable the analysis of sperm chromosomes for gross structural abnormalities and point mutations as well as their ability to fertilize eggs. Such analyses become especially important when space crews become exposed to the radiation environment of deep space for prolonged time periods. Finally, recent studies in environmental toxicology have identified a number of strategies and endpoints for monitoring men and women for potential alterations in reproductive status. In addition, the effects of sex steroids on memory, aggression, and sleep are of increasing interest. Psychological ramifications of reproductive function may now be studied in relation to reproductive hormone function.

Major Goals

The major scientific goal is to characterize effects of the space environment on human reproductive function; more generally, effects on mammalian reproductive function. This includes characterization of stress and the microgravity environment as evidenced by hormonal and physiological alterations, including possible hyperestrogenism to bone loss and behavioral aspects of gonadal failure (see Chapter 4, Behavior and Human Performance). In addition, it is desirable to obtain direct evidence on the effects of the space environment on male and female gametes including the effects of radiation on gametes.

Progress

The Soviet COSMOS flights using rats have provided virtually all of the information that exists on mammalian reproduction in space. The first study was a 19-day flight during which rats were observed for mating and parturition in space. None of the females delivered offspring, which was hypothesized to result from fetal absorption caused by stress. However, there was no reported verification that mating had occurred. During a more recent flight an experiment using rats that became pregnant prior to flight successfully delivered live young in space. NASA has proposed a study, using mice, that will examine mouse embryos prior to implantation and will study their ability to develop in space.

Concerning humans, a number of factors have been identified as having the potential to affect reproduction in space. Some of these come from studies made of men and women undergoing military training or strenuous athletic activities.

Lack of Progress

Aside from the proposed experiment using mice, no additional studies in space have been indicated. There are no indications that NASA intends to monitor the reproductive status of crew members or that consideration has been given to issues of the potential effects of the space environment (microgravity and radiation) on human reproduction.

6

Plant Biology

Plants may be regarded as the most important organisms on our planet as they are key to the entire biological system that has developed on it. It is essential to understand the effects of gravity and its absence in order to employ this knowledge to growing plants in space as well as to enhance their growth and use on Earth. However, our knowledge concerning the growth of plants in the microgravity environment of space is minimal. Of tremendous practical importance is the simple question of whether or not plants can grow and reproduce with normal efficiency through multiple generations in space. Answering this question will require long-term flight experiments together with carefully designed on-orbit and ground controls.

Recommendations and objectives contained in the three SSB/CSBM reports being considered in this assessment remain largely unfulfilled. The 1979 report calls for studies on how plant cell polarity first develops and emphasizes the use of horizontal rotation as a substitute for microgravity. The 1987 strategy is a more comprehensive statement of experimental goals and recommendations related to the growth, development, and behavior of plants in space. The 1988 report questions the threshold of gravity-sensing systems in plants—arguing that without a low-g environment for longer flight periods, resolution of thresholds and response mechanisms cannot be reliably determined. In fact, the most fundamental research requirements to addressing these issues are a "quiet" on-orbit environment on long-term flights and the use of an on-board centrifuge for threshold and control.

STATUS OF DISCIPLINE

It has been demonstrated repeatedly that plants will *grow* under microgravity conditions. However, there are no quantitative studies regarding the efficiencies of growth processes. It remains to be demonstrated that seeds produced in space are capable of *completing a life cycle and producing another generation of normal plants.*

For decades, the clinostat (a horizontal rotation device) has been thought of as a means of reducing the effects of the gravitational vector and thus, in the minds of many, a substitute for microgravity. However, recent studies on-orbit have shown that there is an absence of thermal convection in space which results in a lack of movement of materials and mixing of cell contents. Clinostats exaggerate such mixing and have, in fact, been shown to advance plant cell death as a result. Thus long-term studies of plant responses in clinostats can no longer be viewed as equivalent to experiments in microgravity. The results of the two types of study may sometimes appear superficially to be the same, but will have been derived from virtually opposite physical effects.

Since 1980, investigators supported by NASA have published 335 papers about plants and space biology in reviewed scientific literature. Most have dealt with plant hormones and gravitational responses at the levels of cells, organs, and whole plants. There have also been many phenomenological studies concerned with movement and orientation. Our understanding of plant signal transduction remains scant. However, such constituents as G-proteins, phosphoinositides, actin, and calmodulin also occur in plant cells and may have active roles.

In addition, since 1987, great strides have been made in plant molecular biology. Many of these advances can be applied to problems in plant growth and development, and will be useful in trying to understand plant responses to the space environment. Of particular relevance are advances in our understanding of signal transduction mechanisms. In the last few years there has been a great deal of elegant work on receptor molecules and signal transduction mechanisms for a variety of responses to hormonal and environmental stimuli in animal cells, bacteria, and yeast. Physiological, biochemical, molecular, and genetic techniques are being exploited, with the most significant advances coming from combinations of these approaches. Some investigators are now beginning to apply these techniques to signal transduction in plants, and it is clear that much can be learned.

Particularly promising are recent experiments with auxin receptors and mutants with altered geotropic responses. It is reasonable to suppose that cloned genes for receptors will be available soon, and that genetic analysis of gravitropism will lead to an eventual understanding at the molecular level. Such experiments will also provide characterized genetic material, which may be used in future flight experiments to determine whether genes involved in sensing or responding to gravity on Earth will affect plant growth under microgravity conditions.

MAJOR GOALS

The Goldberg Strategy calls for understanding the effects of gravity or its absence on broad categories of plant responses and processes. Here, building on the Goldberg Strategy with issues raised in other reports, the committee has taken a somewhat "global" approach in detailing major goals for plant reproduction, growth, and development in space, which will have overlapping criteria and priorities, rather than address all of the specific proposals suggested in the Goldberg Strategy.

Experimentation is needed to identify those impediments to sexual and asexual reproduction that have been suggested in some space studies. The single biggest practical question is whether plants are capable of multiple generations in microgravity. Seeds germinated on-orbit would have ground-born flowers and thus will produce partially with ground-born seeds (since seeds are composed in part of material tissues). Only second generation seeds would produce flowers and seeds from tissues exposed only to microgravity. So, in the end, only plants produced from a third planting of seeds would be entirely free of any prior gravitational influence. Thus, the definitive experiment will not be a seed-to-seed experiment, but a seed-to-seed-to-seed experiment.

Fundamentally important experimentation in plant development must include the range from cell to organism and must account for interactions of microgravity with other environmental factors. It has not been clearly shown whether microgravity affects all cells or if some cell types acclimate to gravity deprivation. Some space studies suggest that chromosome behavior is fundamentally changed in weightlessness. If so, what are the consequences for cell development? Biological clocks are important regulators of plant development, but it is not known if they function reliably in space. Radiation in space may also have major developmental consequences and may interact with microgravity effects. For example, mechanisms for repairing radiation-induced genetic damage may be affected by microgravity.

The lack of thermal convection in space may affect many aspects of short- and long-distance transport phenomena in plants. For example, the functions of cell membranes, the pathways for ion uptake and nutrient absorption, water relations, and the transport of organic and inorganic molecules must be investigated with regard to any ways by which they are affected by weightlessness. Some spaceflight results indicate that microgravity affects metabolic products in other ways as well. For example, photosynthesis and pathways of carbohydrate and lipid affect plant growth and/or nutritional quality. Also, protein metabolism may be affected by microgravity. Supporting structures of lignin and cellulose may be modified in ways analogous to the loss of bone density in space. A major goal should be to gain an appreciation for the magnitude of such effects and their impact on plant function.

The need for extensive study of plant gravitational responses has been identified in all previous reports. Because plants are normally stationary, they are more suitable than animals for studies of gravity perception. Many of these experiments must be done in space, since ground-based horizontal rotation techniques do not properly mimic microgravity. At present, the most pressing need is for a definition of the minimum gravitational fields required for responses at the molecular, cellular, and whole plant levels. As emphasized in the Goldberg Strategy, such experiments will require the use of an on-orbit centrifuge to provide various levels of partial gravity levels and normal gravity on-orbit controls.

PROGRESS

Fragmentary Soviet experimentation involving growing plants (*Arabidopsis*) to the flowering stage in space indicates that reproductive events are most affected by microgravity; fruit set is decreased over ground controls and percent seed germination back on Earth is also below control levels.

Experiments studying single cells on the D-1 Mission indicate both stimulatory and inhibitory effects of spaceflight on growth and development. Of note were the observations that the single cell green alga, *Chamydamonas*, maintained a well-expressed circadian rhythm, which did not differ significantly from ground controls, over a 6-day period in space. Cultured cells of the plant Anise turned into embryoids and developed leaf and root primordia and chlorophyll faster in space than did ground-based controls. Observations of cell division in microgravity indicate a generally slower cycle, and several observations of the mitotic sequence indicate abnormalities in chromosome behavior. Other studies have indicated that there are gravity-induced changes in the distribution, and very likely the synthesis, of cellulose (a structural counter to gravity on Earth) and starch (which makes up the gravity sensing statolith (amyloplast)). Added to the latter observation is the increase in numbers and size of lipid bodies in the cells where the amount of starch is decreased.

There seems little doubt left that the amyloplast is the gravity sensor on Earth. On the D-1 Mission, on-orbit seedling roots show no evidence of amyloplast sedimentation, and roots simply grow straight out from their position in the seed.

On-orbit 1-g centrifuge controls behaved as if on Earth. Ground-based investigations have indicated calcium, auxin (a plant hormone), and membrane transport processes in gravity perception. Although much more needs to be done before a satisfactory general model can be formulated, specific hypotheses can now be tested. For example, one can ask the question of whether the sedimentation of starch grains causes changes in ion transport at the cell membrane, and if so, whether such changes are required for gravity sensing.

LACK OF PROGRESS

To date, flight experiments have indicated a variety of differences between the behavior of plants in microgravity and those on Earth. However, both a lack of flight opportunities and the inability to carry out on-orbit centrifuge controls have severely restricted progress in basic plant biology. As a result, we lack information on both direct and indirect effects of microgravity. The magnitude of long-term microgravity effects on such fundamental processes as cell growth and chromosome behavior remains unknown. It is also unclear to what extent changes in convection, surface tension, cohesion, and so on, may influence photosynthetic gas exchange or the transport of water and nutrients.

Long-term flight experiments are necessary to establish if plants grow or can be made to grow under conditions of near-weightlessness. Can they go through a life cycle? Several life cycles? We simply do not know. At present we can only speculate about possible problems. Until there are long-term flight data on plant performance, such questions will remain unanswered.

CONCLUSIONS

In addition to basic scientific curiosity, the desire of mankind for a long-term presence in space is now driving researchers to learn how to grow plants in space as the fundamental components of bioregenerative life support systems. (For a more detailed discussion of this topic, see Chapter 7, Closed Ecological Life Support Systems.) To compress the photosynthetic life support system that sustains all life on Earth into small spacecraft or even a lunar base is a formidable challenge that will require a much more precise understanding of plant capabilities and performance than has ever before been necessary. Plants will need to grow at maximum yields, in exotic environments, with unfailing reliability to play a role in space biohabitats. An unprecedented goal of coupling engineering to plant production will demand reproducible data about plant performance in space and maximum rigorous scientific precision.

In order to approach the goals mentioned above, and those previously emphasized in the 1987 and 1988 CSBM reports, the discipline of space plant biology must have available more frequent flight opportunities, much longer microgravity exposure, and on-orbit controls. Only adequate, replicated, and controlled experiments contribute to science. A series of unrelated anecdotal information does not.

7

Closed Ecological Life Support Systems

The closed ecological life support system (CELSS) program at NASA is attempting to create an integrated self-sustaining system capable of providing food, potable water, and a breathable atmosphere for space crews on long-term missions. CELSS research activities are currently under way at Kennedy Space Center, Ames Research Center (ARC), Stennis, and Jet Propulsion Laboratory. Laboratory research and calculations indicate that average needs for food, water, and oxygen could be met by a "bioregenerative" system utilizing higher plants and/or algae. It is possible that such a system could operate in a small enough volume to be practical in a space vehicle. However, lunar and planetary bases may also make extensive use of a variety of CELSS installations, and in these situations there may be fewer volume constraints.

An effective CELSS must have subsystems for plant growth, food processing, and waste management. Plant growth is expected to occur under controlled conditions with respect to light, temperature, nutrient supply, and atmospheric composition, and must produce the highest practical yields of edible biomass. Optimization of this system will depend greatly on the choice (and perhaps on the breeding) of suitable crop species and varieties. The food processing system must be designed to derive maximum edible content from all plant parts, while the waste management system must recycle the solid, liquid, and gaseous components necessary to support life. Additional concerns arise from the potential for serious problems associated with trace gases and pathogenic organisms in a closed system, and from the necessarily small size of the system and consequent lack of ecological buffering.

Although CELSS is easily articulated at the conceptual level, numerous areas of ignorance remain to be resolved before a safe and reliable system can be achieved. In some cases our ignorance is fundamental and reflects our inability to experiment with reduced gravity environments on Earth. For example, a critical question is whether plants can grow and reproduce well enough to make long-term crop production practical in a reduced gravity environment. No information is currently available on long-term plant growth in reduced gravity; nor can we predict the extent to which reduced gravity may affect such component processes as phloem transport, water movement, gas exchange, photosynthesis and respiration, cell division and expansion, the transition to flowering, fruit[1] set, and seed development.

Other fundamental questions do not necessarily center on reduced gravity environments but on the complexity of closed biological systems. Without extensive experience with such systems, it is difficult to anticipate problems that may arise from the accumulation of trace gases, for example, or from the presence of plant pathogens or viruses affecting microbial species associated with the plant growth or waste management systems.

The closed ecological life support system is much more than a "greenhouse in space." It is a multispecific ecosystem operating in a small closed environment. Even the best systems will have several orders of magnitude less buffering capacity than a normal agricultural environment and will therefore need careful management to provide the ecological stability required in a reliable system.

STATUS OF DISCIPLINE

The approach to CELSS by NASA appears to be from an engineering perspective, without recognition that relevant biological knowledge is missing, and that in the absence of this knowledge, it is impossible to specify reasonable design constraints. CELSS must take into account not only serious questions about reduced gravity environments, but also the many other fundamental questions concerning the operation of complex closed environmental systems. Such systems are comparable in complexity to the human body, but have received only a tiny fraction of research attention to date. We have little experimental data on the actual operation of such systems—what little data we do have come from the Soviets rather than from our own experiments.

Indeed, we have very little data on the operation of individual system components under realistic conditions. A small amount of information has been gathered concerning the performance of a few plant species in open growth chambers, and some encouraging but still very tentative experiments

---

[1]Fruit is a botanical term including food items commonly referred to as *seeds*, such as grains and legumes.

have been initiated on plant growth in a closed environment. Virtually nothing has been done with respect to microbial and other systems for waste recycling; soil and epiphytic microflora; viral, bacterial, and fungal pathogens (various of which may affect plants, synergistic microbes, animals, or humans); or any of the food processing technologies required for converting biomass into palatable human nutrition.

From even such a brief listing as this, it is apparent that CELSS is not a single discipline but an amalgamation of projects in many disciplines. In each discipline, a great deal of scientific information must be gathered before the process of providing engineering design constraints can begin. Both basic and applied information is required to address such fundamental problems as the extent to which growth and reproduction of plants and microbes are affected by reduced gravity, as well as practical problems of, for example food processing and water recycling.

The closed ecological life support system, like other biological research programs, will benefit from advances in genetic engineering technology. Potential applications of this technology to CELSS range from the use of molecular markers to accelerated plant breeding programs to techniques for the directed transfer of isolated single genes. Molecular probes may also be useful in monitoring populations of microbial pathogens.

To exploit the opportunities presented by this new technology and to address the many remaining uncertainties concerning growth, development, and reproduction in a space environment, it is essential for the CELSS program to progress simultaneously on many fronts. It must also retain the flexibility to respond quickly and creatively to increases in scientific knowledge or to unanticipated biological problems. As noted in the 1988 Life Sciences Task Group report, the integrative multidisciplinary nature of the CELSS program should allow it to serve as a focal point for many other basic and applied research programs. Considerable synergism would be expected from such a relationship.

MAJOR GOALS

Previously listed goals range from extremely general to quite specific. The 1979 report discussed CELSS largely in the context of issues related to system closure. Experiments on the closure of natural ecosystems were recommended because work with such systems might identify unanticipated problems relevant to the performance of agriculturally oriented systems. In addition, the report emphasized the importance of a broad survey of plant and animal species for use in CELSS. Cooperative efforts by anthropologists, ecologists, nutritionists, and agricultural scientists were recommended, and it was emphasized that the species considered should not be restricted to those of conventional terrestrial agriculture.

The closed ecological life support system was not considered as such in the Goldberg Strategy, but many of the critical issues identified in that report apply with force to any program involving living organisms. Notably, this report highlights our ignorance concerning the effects of radiation and reduced gravity on the long term performance of organisms in space. A major scientific goal articulated in the 1987 report was to "evaluate the capacity of diverse organisms, both plant and animal, to undergo normal development from fertilization through the subsequent formation of gametes under conditions of the space environment." For CELSS, this overall scientific goal assumes immediate practical importance. Even small deleterious effects on growth and reproduction may have serious consequences for processes such as food production in which crop performance is integrated over long time periods that may involve several reproductive cycles.

The 1988 report listed a number of specific goals for the CELSS program. Among them were determining environmental requirements for higher plants in closed systems, evaluating the use of algae as potential food sources, and research on biological waste processing. That report highlighted the formidable complexity of the CELSS task and the opportunity for CELSS to become a focal point for much of NASA's biological research.

PROGRESS

From an initial consideration of primarily agricultural species, a small number of plant species have been selected for further investigation. These include wheat, potato, soybean, and tomato. Growth chamber studies have been initiated, both in NASA and in university laboratories, with the aim of defining environmental conditions and plant nutritional requirements for optimum rates of dry matter production. Although most of this work is being done with open systems, experiments with a closed chamber have recently been initiated.

The first experiments have been designed to test the effects of atmospheric closure on plant performance and have not focused on water and waste recycling. However, these issues can be addressed as the program develops.

A closed ecological life support system in space will require specially designed systems capable of supplying water and dissolved nutrients to plant root systems. Conventional hydroponic systems cannot be used effectively in microgravity, although they have many advantages on Earth. However, NASA investigators are developing a nutrient delivery system for this purpose. The system has been tested successfully in the CELSS chamber at Kennedy Space Center although, so far, there has not been an opportunity to test it in space.

Some effort has been devoted to assessing intra-specific variability in performance under controlled environments, mainly involving growth chamber tests on a large number of varieties of wheat. The results have been instructive, since there is a high degree of genetic variation and only a few of the many lines studied approach optimum performance levels under the test conditions used. This result highlights the importance of variation and the need to examine many genetic stocks before making decisions on the suitability of a given species for CELSS. Perhaps even more importantly, the existence of such extensive genetic variation might be exploited in future plant breeding efforts. Plant breeding programs have tremendously improved the productivity of terrestrial agriculture, and similar techniques could undoubtedly be used to improve the performance of selected plants in closed systems.

LACK OF PROGRESS

Thus far, progress in CELSS has been constrained by low funding levels, which have contributed to an overly narrow focus. In addition, the program is severely handicapped by a lack of information concerning long-term plant growth in space.

Ground-Based Research

Research areas in which a great deal of work can be done in ground-based laboratories include the following. (1) Plant growth in controlled environments has received attention in the past, but there is still a need for much more extensive research in this area. (2) A facility is required in which many different combinations of environmental variables can be tested in parallel. (3) The use of ambient light in addition to internally generated light needs investigation for use in situations (such as a lunar CELSS) where sunlight is available. (4) Attention should be paid to the effect of diurnal and other rhythms that may affect plant growth. (5) Many more plant species need to be tested. All such experiments, up to and including those at the KSC Breadboard project level, should be conducted with simultaneous controls so that the effects of experimental variables can be evaluated more quickly and reliably. (6) Higher photosynthetic efficiencies can be achieved with algal systems than with land plants, but recent studies have not emphasized algae because of perceived difficulties in preparing them for use as food. This matter should be investigated further in close collaboration with food scientists. (7) Food processing has received relatively little attention so far. A much greater effort is justified, and there should be close coordination between food processing and plant production research activities. (8) Waste processing systems have also received surprisingly little attention in view of their crucial importance. A wide range of extensively tested options should be available for use in developing fully functional CELSS systems. (9) Plant diseases and bacterial viruses have hardly been studied at all in closed systems, but knowledge of such organisms will be crucial. We must understand the epidemiology of such infections through a program of carefully controlled experiments; anecdotal accounts will not have the required predictive value.

Only a very limited number of species have so far received consideration for use in CELSS—testing biological diversity should be a fundamental goal. In addition, although it has been shown that wheat varieties differ dramatically in their responses under conditions currently deemed appropriate for a CELSS, no systematic plant breeding efforts appear to have been attempted.

Flight Experiments

As noted above and in Chapter 6, Plant Biology, we do not yet know if plants will grow in space sufficiently well to support a CELSS for significant periods of time. Processes such as reproductive development, fluid transport, and photosynthetic gas exchange may be adversely affected in low-gravity environments.

Experiments to determine the severity of effects on these and other processes are conceptually very simple, but depend on the ability to grow plants in space for extended periods of time, preferably several life cycles. A great deal of information useful to CELSS can be obtained from studies with model plant systems such as *Arabidopsis*, provided such studies are thorough and properly controlled. Thus it will be important to extensively interface and coordinate the CELSS and space biology flight programs. However, additional flight experiments will be required to test a number of plant and algal species and to compare their responses to low-gravity space environments.

Both basic and applied plant research programs in microgravity will be most effective if varying levels of artificial gravity are provided in a space-based centrifuge. Extensive use of a properly designed centrifuge facility will help investigators separate the effects of gravitational and other environmental factors and achieve a greater mechanistic understanding of plant responses. Such a mechanistic understanding is required if we are to be able to make predictions concerning plant performance under conditions different from those actually tested.

CONCLUSIONS

Whereas NASA's CELSS research should be focused to address mission-related questions, the program must also develop a much broader base of scientific knowledge and the ability to take a more flexible approach to system design. Expanded basic research and development efforts are required in all areas of the program.

At least in principle, much of the missing information required for CELSS design can be obtained fairly rapidly, and—again, in principle—it is probably reasonable to expect that a prototype CELSS could be tested on the moon. However, meeting these objectives will require a much greater commitment to both basic and applied biological research than NASA has thus far been willing or able to make. Plant, microbial, waste processing, and food technology programs in CELSS, as well as plant and microbial programs in gravitational biology, need to be greatly expanded, and flight opportunities must be provided for suitable, controlled experiments on long-term growth of plants in space.

8

Radiation Biology

The radiation environment of space is considerably less benign than on Earth. Planning for extended human sojourns in space mandates that we have quantitative knowledge about the dose rates and the types of radiation that will be encountered. Similarly, the effects of the different types of radiation encountered in space, especially deep space, must be defined quantitatively. Radiation environments will be defined as a result of dosimetric measurements made in space and models that include attention to factors that cause the marked temporal variations in radiation fluxes. Much of the necessary radiobiology research can be carried out on Earth with defined radiation sources. Basic experiments in space are required to investigate and understand the interactions if any between radiation and microgravity.

STATUS OF DISCIPLINE

The radiation environment within the magnetosphere is known with fair precision. Measurements on Shuttle missions at different altitudes and orbital inclinations have established the ability of models to predict dose rates encountered in low earth orbits. Still, the precision of the information about the radiation environment that will be experienced in geosynchronous orbit and within space vehicles in deep space is not yet adequate for the estimation of risks of radiation effects.

The biological effects of low-energy transfer (LET) radiations, such as protons and electrons, are relatively well known because of the large body of data obtained from ground-based studies. This is not the case for high-energy (HZE) particles that are a small but important component of galactic cosmic rays.

Current knowledge about radiation environments and radiobiology relevant to space has been summarized in a number of reports, including the Space Science Board's 1988 report *Space Science in the Twenty-First Century—Life Sciences*.

SCIENTIFIC GOALS

The Goldberg Strategy did not deal comprehensively or specifically with radiation studies. The report did, however, note some scientific objectives, for example, (1) to measure the specific effects of radiation versus gravity on genomic stability and the appearance of aberrant cell lineage over several generations of representative organisms; (2) to establish an extended program of research into the effects of HZE particles on developmental events; and (3) to investigate whether HZE particles enhance the probability of malignant transformation.

PROGRESS

Despite budgetary restrictions, NASA has managed to maintain a limited but ongoing research programs in the fields of both radiation dosimetry and radiobiology. Studies of radiation environments have been carried out at both the Agency's own centers and several universities. Measurements of radiation of various types have been made on all Shuttle missions as well as some Soviet missions in collaboration with USSR scientists. Instrument development has also been supported. Ground-based studies are currently under way on the effect of fragmentation of HZE particles and on the secondary particles, areas of importance for deep space missions.

In the field of radiobiology, a modest ground-based program has been supported by NASA. The Agency has had to rely mainly on universities and the national laboratories for such research as it does not now have any radiobiological studies at its own centers. The radiobiological studies supported by NASA and other agencies have been concerned mainly with the effect of HZE particles on DNA, cell survival, mutation, malignant transformation of cells *in vitro*, carcinogenesis, and the induction of cataracts and life shortening in mice. These studies have provided some information about the relative biological effectiveness (RBE) of a number of different types of heavy ions. Certain generalizations about the relationship of LET and RBE for some specific acute and late effects can be made from these studies. For example, the values for RBE increase with increasing LET up to about 100-200 KeV/μm.

Studies were carried out on the German D-1 Mission to determine whether there were interactions between radiation and microgravity on embryogenesis and organogenesis in C. morosus. Results indicate that (1) HZE particles caused anomalies when traversal of HZE particles occurred during organogenesis and (2) the combined exposure to HZE particles and microgravity acts synergistically on eggs at all developmental stages, resulting in high rates of developmental anomalies.

Recently, NASA supported a study by the National Council on Radiation Protection (NCRP) of the available information in order to estimate the risks from radiation exposures in space. This study has resulted in the development of guidelines concerning radiation protection standards that can be applied to astronauts on missions in low earth orbit such as the Space Station.

The radiation environment inside space vehicles in deep space must be defined more precisely. There are plans to make the measurements of radiation fluxes beyond the Van Allen radiation belts and to determine the energy and LET spectra of the various types of radiation. For all the radiation environments in space, there is a need to acquire real-time measurements of the energy, LET spectra, and LET fluxes of the different types of radiation involved.

Continuing ground-based radiobiological studies should (1) determine the effects of protracted proton irradiation (protons predominate in low earth orbits and are the major component of galactic cosmic rays); (2) investigate the effects of HZE particles, including cancer induction; and (3) establish whether HZE particles cause damage that may only appear late in life and that cannot be predicted from knowledge about the effects of other radiation qualities. Basic studies on the relationship of energy deposition and biological effects with HZE particles can provide information important to the general understanding of radiation effects.

There is a need for a number of studies using different biological systems to determine the nature and importance of interactions of radiation and microgravity. These studies should be broad in range including investigations of chromosome stability and chromosome aberrations, cell proliferation, and cellular functions important to immune competence. The studies of the effects of microgravity on embryogenesis and development and growth are covered elsewhere in this report, but these investigations will provide a further opportunity of studying radiation effects and the question of interaction between radiation and microgravity. To undertake studies of the effects both of radiation alone and of interactions between radiation and microgravity, "LifeSat" has been proposed as a candidate for a FY 1992 new start. This free-flying satellite could accommodate various experiments including both rodents and simpler organisms that have been suggested and are under consideration. It is important that the sample sizes in such experiments are sufficiently large to ensure unequivocal results. For most endpoints this will mean that simple organisms be chosen. With experiments that use test systems such as *C. elegans*, it will be possible to study a number of different endpoints, including aspects of development.

As the duration of the missions is extended, it will become possible to study the important aspect of protracted irradiation, and such studies should be carried out.

LACK OF PROGRESS

Because of limited flight opportunities to date, it has not been possible to conduct proposed studies of interactions of radiation and microgravity.

NOTE ABOUT FACILITIES

Current U.S. plans call for the BEVALAC facility at Lawrence Berkeley Laboratory, (the only U.S. facility currently capable of providing beams appropriate for the study of very high-Z particles) to be phased out. It will probably not be available after 1993. A proposed replacement for the BEVALAC would provide heavy-ion beams but not the very high-Z particles, such as iron, which is the heavy ion of particular interest in space radiation studies.

It has been suggested that an alternative high-energy radiation research facility is the synchrotron at Darnstadt. There are, however, a number of practical considerations as to why this would not be desirable: (1) no biological experiments have yet been performed at the synchrotron; (2) there are currently a large number of investigators waiting to use the facility; and (3) there are no existing animal holding facilities, and none are planned. The use of a large number of animals is fundamental to this type of research.

Time is short for providing an alternative research source to study heavy ions in order to continue the HZE-particle studies. If plans for extended human space missions are to be seriously pursued, it is essential to have a facility readily available for the study of HZE particles.

9

Conclusions

As pointed out in the Goldberg Strategy, space biology and medicine is in the earliest stage of discipline development. Relatively few experiments have been flown; most have not been part of a larger research strategy; and few have been well controlled or replicated. Nevertheless, 25 years of observations carried out in space leave little doubt that the microgravity of space results in significant alterations not only in the physiology of organisms but also in the function of individual cells. Some of these changes are potentially life-threatening. Hence, a major concern of research in space biology and medicine is to guarantee the health and safety of the humans who undertake missions in space.

The health and safety issue requires some clinical research that ideally should be conducted in ground-based laboratories as well as in the microgravity environment of space. However, most of the required research could be classified as basic. Instead of developing protocols to treat or mask the various biological changes that occur during spaceflight, a short-term expedient, the CSBM feels strongly that the basic mechanisms underlying microgravity-induced changes must be understood. Studies of the basic mechanisms underlying overt clinical changes seen in spaceflight should have the first priority in any relevant research strategy.

The preceding chapters illustrate that NASA has made an effort to implement parts of the research strategies provided by the Space Studies Board. However, progress has been painfully slow, and it appears that considerable hope remains for a "quick fix." This hoped for success seems doomed to failure. To succeed, both the increase in funding and the marshalling of expertise will have to be exceptional, in amount and alacrity. Most of the Goldberg Strategy, as well as recommendations from other reports, remains valid and awaits implementation. What is required is a commitment to a long-term, sustained effort to understand the nature of the effects of microgravity on living processes at several levels.

There have been and remain two major limitations to implementation of the research strategies previously published. The first is an adequate commitment of resources to space biology and medical research. The proportion of the life sciences budget that has been dedicated to research has been less than the support necessary to support a single moderate-sized department in a major research-oriented university. In view of the enormous breadth of basic and clinical research that must be carried out in the life sciences to ensure the future of human presence in space, the lack of a serious commitment to the life sciences program is not commensurate with agency plans for future human space exploration.

If the nation and NASA are committed to a program of human exploration, a substantial infusion of funds is a prerequisite for success. Part of the funds should be used to enhance ground-based research, not only in areas concerned with human physiology, but especially in research areas concerned with human behavior. A concerted national effort in these areas will require not only NASA participation but also that of other major federal agencies concerned with biological research.

However, even if a considerable infusion of funds into the space biology and medicine program were to occur, it is not clear that NASA has or should have the personnel or facilities to take full advantage of these additional funds. Thus the committee's recommendation that NASA initiate increased interaction with other federal agencies concerned with research in basic biology and medicine becomes paramount. Specifically, the Committee on Space Biology and Medicine suggests that *the Nation direct the relevant federal agencies (i.e., NIH, NSF, USDA) to encourage investigators to undertake ground-based research programs concerned with the major research topics related to the health and welfare of humans in space.* This includes research topics not only in human physiology and behavior but also more basic research in areas such as developmental biology and the ability of plants to grow and reproduce in a microgravity environment. NASA could play a primary role in designing and implementing experiments that confirm and enhance models derived from the ground-based basic research.

The second major impediment to progress has been the lack of access to space. This is an obvious and generic problem common to all the space research disciplines. It must be emphasized that since 1985, no space missions have been flown that were dedicated to research in space biology and medicine, and relatively few are planned prior to the projected completion of Space Station. Considering that the discipline was in its infancy at the time the Goldberg Strategy was published, it is likely to remain so until the Space Station becomes fully functional. In this sense, the provision of more

CONCLUSIONS

research space on Shuttle flights and the utilization of retrievable satellites for basic research in space biology and medicine (LifeSat) are imperatives in the interim until construction of Space Station Freedom is completed.

Related to this, maximizing the design and utility of the Space Station for research in space biology and medicine should receive highest priority. Specifically, there should be *a dedicated life sciences laboratory on space station (not shared with other disciplines) and research on space biology and medicine of the space station should be divided into 3- to 6-month blocks with each block devoted to a single research area, e.g., bone and muscle physiology.*

The latter recommendation is especially important. As discussed in previous sections, research in space biology and medicine is empirical. Scientists in ground-based laboratories design experiments that are frequently modified as results are obtained and new technologies develop. This flexibility is particularly pertinent when one considers that most experiments in the queue for flight in the 1990s were designed in the late 1970s. Current technical approaches employing the most recent and rapid advances in molecular biology were almost nonexistent at that time.

Simple observations related to the rhythms of biological changes were also less clear at that time, e.g., hormone levels change as a function of the time of day. Yet, studies manifested on an upcoming Shuttle mission (SLS-1) will investigate hormonal changes in blood samples taken at a single time point. (See Chapter 4, section on circadian rhythms, and Chapter 3, section on cardiovascular hormones and stress, for a further discussion of this issue.) These kinds of problems can be circumvented with a dedicated block of time in a dedicated research facility in which constant and rapid change in experimental design can be accomplished.

The final point of major concern to the CSBM is the acquisition, handling, and dispersal of data obtained from both ground-based and space research. To date there is no centralized data base on results from previous experiments, nor do there appear to be plans to create one. Thus, the current data base, albeit meager, is not readily available to investigators who might wish to design new experiments. The result of this deficiency is to severely restrict access to the research program in space biology and medicine. At the minimum, existing data should be summarized in peer-reviewed journals readily available to any investigator who wishes to initiate a research program in space biology and medicine. As was mentioned in Chapter 2 regarding data management, the CSBM has been informed that plans and discussions are under way at NASA to establish a data and flight specimen archive following the SLS-1 mission. The committee encourages NASA to do so and to solicit continuing input and advice from this and other committees as the system becomes more well defined.

Ultimately, progress toward implementation of the research strategies in space biology and medicine will require a complete reassessment of the approaches currently in place not only within NASA but by the nation. Otherwise, as stressed in the Goldberg Strategy, "based on what we know today, the assumption of continued success of missions involving the sustained presence of humans in space for months to years at a time cannot be rigorously defended."

Bibliography

Documents on Space Biology and Medicine

Dubois, A.B., et al.(1963) *Respiratory physiology in manned spacecraft*, symposium presented at Annual Meeting of Federation of American Societies for Experimental Biology, April 19, 1963. *Federation Proceedings* 22 (Part 2): 1022-1063.

SSB(1963) *Summary Report on Gaseous Environment for Manned Spaceflight*, National Academy of Sciences, Washington, D.C.

SSB(1966) *Physiology in the Space Environment, Vol. II*, National Academy of Sciences, Washington, D.C.

SSB(1967) *Radiobiological Factors in Manned Space Flight*, Space Radiation Study Panel, Life Sciences Committee, National Academy of Sciences, Washington, D.C.

SSB(1968) *Physiology in the Space Environment, Vol. I*, National Academy of Sciences, Washington, D.C.

SSB(1969a) *Report of the Panel on Management of Spacecraft Solid and Liquid Wastes*, National Academy of Sciences, Washington, D.C.

SSB(1969b) *Report of the Panel on Atmosphere Regeneration*, National Academy of Sciences, Washington, D.C.

SSB(1970a) *Space Biology*, National Academy of Sciences, Washington, D.C.

SSB(1970b) *Radiation Protection Guides and Restraints for Space Mission and Vehicle Design Studies Involving Nuclear Systems*, National Academy of Sciences, Washington, D.C.

SSB(1970c) *Life Sciences in Space*, Study to Review NASA Life Sciences Programs, H.B. Glass, ed., National Academy of Sciences, Washington, D.C.

SSB(1970d) *Infectious Disease in Manned Spaceflight*, National Academy of Sciences, Washington, D.C.
SSB(1973) *HZE-Particle Effects in Manned Spaceflight*, Committee on Space Biology and Medicine, National Academy of Sciences, Washington, D.C.
SSB(1974) *Scientific Uses of the Space Shuttle*, National Academy of Sciences, Washington, D.C.
SSB(1975) *Opportunities and Choices for Space Science 1974*, National Academy of Sciences, Washington D.C.
SSB(1976) *Report on Space Science 1975*, National Academy of Sciences, Washington, D.C.
SSB(1979) *Life Beyond the Earth's Environment*, National Academy of Sciences, Washington, D.C.
SSB(1987) *A Strategy for Space Biology and Medical Sciences for the 1980s and 1990s*, National Academy Press, Washington, D.C.
SSB(1988) *Life Sciences*, volume in a series, *Space Science in the Twenty-First Century*, Task Group on Life Sciences, National Academy Press, Washington, D.C.

Space Studies Board Letter Reports

To. J. Beggs regarding space station centrifuge, September 9, 1983.
To. A. Stofan regarding centrifuge, July 21, 1987.
To. R. Truly regarding extended duration orbiter medical program, December 20, 1989.
To. J. Alexander regarding Space Station Freedom pressurization and atmosphere, December 12, 1990.
To. R. Truly regarding space station position on proposed redesign of Space Station Freedom, March 14, 1991.

Appendix

Guidelines for Assessment Reports for Standing Committees of the Space Studies Board

So that the Space Studies Board (SSB) can have an ongoing assessment of the status of space science and applications research recommended in its various reports, each of the standing committees (Space Biology and Medicine, Space Astronomy and Astrophysics, Solar and Space Physics, Planetary and Lunar Exploration, Microgravity Research, and Earth Studies) is requested to provide an assessment of the way in which recommendations in the existing strategy and other reports are being implemented by the appropriate federal agencies. This assessment will be conducted every three years beginning in 1990-91. In the interim years, the committee chairpersons will provide a formal assessment report to the SSB only. The form of report presentation, written or oral, is at the discretion of the committee. Should the SSB determine that the reports' contents or format needs to be changed, the SSB will provide the committees with the necessary guidelines to make the appropriate modifications.

A secondary objective of these assessments is for the committees to examine their existing strategies to determine if any changes are necessary and to evaluate the time scale on which the strategies need to be updated. This report is to be submitted to the SSB for review no later than March 31 of the pertinent year. Its length should be determined by the committee based on its individual needs. Each report should include an executive summary, not to exceed 15 pages.

The audience for these reports is NASA, the space research community, Congress, and relevant Executive Branch offices such as the Office of Science and Technology Policy, the Office of Management and Budget, and the National Space Council.

These reports will be published separately, as they become available. In addition, the SSB may choose to summarize and compile all of the reports into a single volume providing an overall assessment of the state of space research on a regular basis.

Each report should contain, at a minimum, the following features:

I. Introduction

A. Description of principal areas/disciplines within committee's purview.

B. Listing of principal SSB reports pertaining to the respective disciplines (should include title, name of authoring committee, date of publication). SSB or committee letter reports that contain information relevant to issues examined in the annual report should also be included. [In some cases, there may be non-SSB NRC reports for which the committee has oversight reponsibility. In these cases, they too should be listed.] This listing may be attached as an appendix.

C. Identification of principal users/implementors of existing reports (within NASA).

D. Identification of potential users/implementors of existing reports (outside of NASA).

II. Status of the Discipline

A. Discussion of *major* scientific goals/objectives in existing reports, and description of progress to date in achieving these goals/objectives.

"Progress" includes any activity specifically pertaining to major scientific objectives, i.e., inclusion in a strategic plan, Announcements of Opportunity, cooperative agreements, budget line-items, inclusion on STS/ELV flight manifest, Phase A, B, etc., studies.

B. Identification of *major* scientific goals/objectives in existing reports in which *no* progress (see above) has been made.

This discussion should include the committee's assessment of why no progress has been made in addressing scientific goals/objectives (e.g., budget constraints, lack of flight opportunities, technology limitations, instrument/facility availability, management/policy decisions, and so on). It should also address, if applicable, where these goals or objectives fall in the committee's overall priorities for the discipline as a whole.

If relevant or possible, the committee should include some guidance or recommendations for achieving these goals and objectives. For example, the committee might recommend how relevant federal agencies (other than NASA) could address these goals and objectives under existing programs, how interagency and/or international cooperative agreements might be exploited, and so on.

C. Identification and discussion of major policy and program issues raised in existing reports and the U.S. government's and/or NASA's response.

III. Conclusions

www.ingramcontent.com/pod-product-compliance
Lightning Source LLC
Chambersburg PA
CBHW081735170526
45167CB00009B/3824